EDITORIAL BOARD

ROBERT C. CLARK
Directing Editor
Distinguished Service Professor and Austin Wakeman Scott
Professor of Law and Former Dean of the Law School
Harvard University

DANIEL A. FARBER
Sho Sato Professor of Law and Director, Environmental Law Program
University of California at Berkeley

HEATHER K. GERKEN
J. Skelly Wright Professor of Law
Yale University

HERMA HILL KAY
Barbara Nachtrieb Armstrong Professor of Law and
Former Dean of the School of Law
University of California at Berkeley

HAROLD HONGJU KOH
Sterling Professor of International Law and
Former Dean of the Law School
Yale University

SAMUEL ISSACHAROFF
Bonnie and Richard Reiss Professor of Constitutional Law
New York University

SAUL LEVMORE
William B. Graham Distinguished Service Professor of Law and
Former Dean of the Law School
University of Chicago

THOMAS W. MERRILL
Charles Evans Hughes Professor of Law
Columbia University

ROBERT L. RABIN
A. Calder Mackay Professor of Law
Stanford University

CAROL M. ROSE
Gordon Bradford Tweedy Professor Emeritus of Law and Organization and
Professorial Lecturer in Law
Yale University
Lohse Chair in Water and Natural Resources
University of Arizona

ENERGY LAW

ALEXANDRA B. KLASS
Distinguished McKnight University Professor
University of Minnesota Law School

HANNAH J. WISEMAN
Attorneys' Title Professor of Law
Florida State University
College of Law

CONCEPTS AND INSIGHTS SERIES®

The publisher is not engaged in rendering legal or other professional advice, and this publication is not a substitute for the advice of an attorney. If you require legal or other expert advice, you should seek the services of a competent attorney or other professional.

Concepts and Insights Series is a trademark registered in the U.S. Patent and Trademark Office.

© 2017 LEG, Inc. d/b/a West Academic
    444 Cedar Street, Suite 700
    St. Paul, MN 55101
    1-877-888-1330

Printed in the United States of America

ISBN: 978-1-63460-290-7

TABLE OF CONTENTS

Introduction 1
I. The Field of Energy Law and This Publication 1
II. Energy Resources and Their Uses 3
III. Federal and State Laws Governing Energy Systems and Industries 4
IV. Federal and State Agencies Regulating the Energy Industry 5
V. Basic Administrative Law Concepts 7

Chapter 1. Energy Facility Siting 13
I. Introduction 13
II. Energy Generation Facilities 13
III. Electric Transmission Lines 18
IV. Oil and Gas Pipelines 22
    A. Oil Pipelines 22
    B. Natural Gas Pipelines 27
V. LNG Terminals 28
VI. Coal Transport: Rail and Coal Export Facilities 32
    A. U.S. Coal Export Trends 32
    B. U.S. Coal Transport and Regulation of Coal Export Facilities 34

Chapter 2. Oil and Gas Extraction 41
I. Introduction 41
II. Division of State and Federal Authority over Oil and Gas Development 42
III. Onshore Mineral Ownership and Leasing 46
    A. Identifying Types of Mineral Ownership 47
    B. Oil and Gas Conservation Regulation 51
    C. Leasing of Privately-Owned Minerals 53
    D. Surface and Mineral Owner Disputes 54
    E. Leasing of Onshore Publicly-Owned Minerals 55
IV. Offshore Oil and Gas Leasing and Regulation 57
V. Unconventional Onshore Mineral Extraction 61
    A. Regulation of Unconventional Onshore Oil and Gas Development 63
    B. The Common Law and Hydraulic Fracturing 70
    C. Private Law Approaches to Controlling the Environmental Impacts of Hydraulic Fracturing 71

Chapter 3. Energy Transportation: Accessibility, Reliability, and Cybersecurity 73
I. Introduction 73
II. Regulation of Electricity Transmission 73
    A. Transmission Access 74
    B. Transmission Cost Allocation 79
    C. Transmission Interconnection 81
III. Safety and Cybersecurity Issues in Energy Transportation 85
    A. The Electric Grid 86
    B. Pipelines 90
    C. Rail 94

TABLE OF CONTENTS

Chapter 4. Environmental Regulation of Energy Production and Use 97
I. Introduction 97
II. Environmental Regulation of Coal Extraction 97
III. Environmental Regulation of Oil and Gas Operations 105
IV. Reducing Air Pollution from the Transportation Sector 110
    A. Regulation of Vehicle Emissions 111
    B. Regulation and Use of Biofuels 113
V. Environmental Regulation of Air Pollution 119
    A. Federal Environmental Laws and Regulations Governing the Electric Power Sector 119
    B. State Environmental Laws Governing the Electric Power Sector 127
    C. Regional Collaborations to Limit GHG Emissions from the Electric Power Sector 129
VI. Environmental Regulation of Hydropower 130
VII. Regulation of Nuclear Power 134

Chapter 5. Energy Financing, Incentives, and Mandates 139
I. Introduction 139
II. Tax Incentives for Energy Production and Use 140
    A. Fossil Fuel Tax Credits and Incentives 141
    B. Major Renewable Energy Tax Credits and Incentives 143
III. Energy Mandates 149
    A. Mandates and Minimum Payments for Renewable Electricity 149
    B. Mandates for Biofuels 155
IV. Energy Financing 158
    A. Financing for Utility-Scale Renewable Energy Infrastructure 158
    B. Financing for Distributed Renewable Energy Infrastructure 161

Chapter 6. Energy Markets 165
I. Introduction 165
II. Federal Regulation of Energy Markets 167
    A. Federal Regulation of Electricity Markets 169
    B. Federal Regulation of Natural Gas Markets 181
    C. Federal Regulation of Oil Markets 184
III. State Regulation of Energy Markets 187
    A. State Regulation of Electricity Generation for Retail Sale 187
    B. State Regulation of Intrastate Transmission Lines 188
    C. State Regulation of the Distribution of Electricity to Retail Customers 190
    D. Oil and Natural Gas: State Regulation of Intrastate Pipelines, Distribution Lines, and Retail Sales 191
    E. Conflicts Between State and Federal Jurisdiction 192

TABLE OF CONTENTS

Chapter 7. Energy Imports and Exports 197
I.   Introduction 197
II. Energy Import and Export Trends 198
III. U.S. Regulation of Energy Imports and Exports 199
     A. Crude Oil Exports 200
     B. Liquefied Natural Gas Exports 202

Chapter 8. The Smart Grid and Distributed Generation 209
I.   Introduction 209
II. The "Smart Grid" 211
III. Electric Vehicles (EVs) 220
IV. Energy Storage 225
V.   Microgrids 228

TABLE OF CASES 233

INDEX 235

ENERGY LAW

INTRODUCTION

I. The Field of Energy Law and This Publication

Energy law is an exciting and developing field that brings together the physical world of natural resources with the technology of energy production, the use of energy in every aspect of our daily lives, and the economic and environmental regulation of energy resources. The goals of this legal field are to promote production, ensure access at a reasonable cost, and minimize the environmental externalities that accompany most forms of energy production, transport, and use.

Energy law in its most basic form is as old as the United States itself, with the common law of property rights, nuisance, trespass, and waste governing the production and ownership of early oil, gas, coal, and hard rock mineral resources. After the invention of the electric light bulb, state utility regulation laws and ultimately, the Federal Power Act, heavily regulated the electric industry. This regulation was designed to ensure access to and reasonable rates charged for this new, important energy resource in the early 20th century. Regulation of natural gas resources at the state and federal levels developed during this same time period. With the invention of the automobile, domestic and foreign oil resources took on outsized importance and, in the 1970s, led to a recognition that a more comprehensive national energy policy was critical to ensure a continued supply of petroleum at a time when the Arab Oil Embargo and diminished U.S. domestic oil production posed a threat to our energy security and modern way of life. As our dependence on fossil fuel resources grew throughout the 20th century, the United States also developed a new energy resource—nuclear energy. While nuclear energy revolutionized the U.S. electricity sector, it also brought new concerns over its use in nuclear weapons, the risk of nuclear meltdowns and other disasters, its growing cost, and the continuing problem of how to address nuclear waste disposal.

By the 1970s, the growing environmental movement brought renewed focus on the air, water, and other pollution that accompanies the production, transport, use, and disposal of energy resources. The development of the federal Clean Air Act, Clean Water Act, National Environmental Policy Act, hazardous waste laws, and a wide range of other federal environmental laws and state laws now has a major impact on the energy law field and our views toward different types of energy resources. The 1990s brought an increased focus on the problem of climate change and the role of fossil fuel use in the energy sector as a major contributor to global climate change.

This led to more tax incentives and mandates for increased use of renewable energy—particularly wind and solar energy—over the next decades. Many believe that renewable energy has the potential to completely transform our energy sector once again, just as oil, gas, and nuclear energy did decades ago. Thus, energy law encompasses the traditional regulation of electricity, natural gas, nuclear, and other energy resources as well as natural resources law, property law, environmental law, and tax and finance law.

Throughout this book, you will see familiar themes arise in multiple chapters, so we take the opportunity to highlight them here. First, as is likely evident to the newest student of energy law, energy law is big business. A quick review of the U.S. and world's largest companies reveals many oil and gas and other energy related companies. Thus, a continuing tension within energy law is the extent to which free markets should govern energy decisions and the extent to which government regulation is needed to guide those markets to ensure safe and affordable access to energy by the public, to reduce (or at least regulate) monopoly power, and to ensure a fair playing field for market participants.

This book is full of examples of policymakers responding to perceived problems in energy markets through new laws and regulations. This raises a second theme that is prominent in this book—if regulation is necessary, what level or levels of government should do it? You will see that many areas of energy law are regulated at the federal level, and others at the state and local levels, and that there are often tensions between federal, state, and local policymakers as their attempts to regulate the energy industry overlap and, in some cases, conflict. For instance, in recent years, the widespread use of hydraulic fracturing technologies to extract oil and gas resources from shale rock has resulted in cities and towns enacting new zoning ordinances to ban or limit the practice or to place restrictions on the time, place, and manner of production. This has prompted several states, such as Texas and Oklahoma, to place new limits on the authority of cities and town to enact such ordinances. Courts are called upon to resolve many of these types of intergovernmental conflicts, and they look to the Supremacy Clause of the U.S. Constitution, federalism theory, local land use authority (in the case of state-local conflicts), and principles of statutory interpretation to do so.

Finally, the tension between energy production and environmental protection looms large whether the topic is oil and gas development, nuclear energy, the continued use of fossil fuels throughout the energy system, and the pros and cons of different types of renewable energy development. You will see that environmental protection provisions are built in to energy and natural resources law directly in some instances, and these provisions also act as a separate overlay in others.

We envision this book as a resource for a variety of audiences. First, teachers of undergraduate, graduate-level, and law school energy law and energy policy courses can use it as primary text, with or without supplemental materials. Second, law school instructors who wish to use a traditional textbook with lengthy excerpts of judicial decisions can use this book as a supplement for students for basic information on many of the topics covered in a traditional law textbook. Third, energy law practitioners, judges, and regulators can use the book as a helpful desk reference for background on a wide variety of energy law and policy topics. We hope that you find the book useful, and we welcome your comments on it.

II. Energy Resources and Their Uses

The primary purpose of energy law is to promote and regulate the energy resources that we rely on for virtually every aspect of our lives. These resources are discussed in detail in the chapters that follow, but a few points are worth noting at this stage. First, energy resources can be roughly divided into renewable and non-renewable resources. Nonrenewable resources are fossil fuel resources—coal, oil, and natural gas—as well as uranium, which is mined from the ground and used to generate nuclear energy. Nonrenewable resources in general and fossil fuels resources in particular have dominated our energy system for decades. Renewable energy resources include wind, solar, hydropower, geothermal, and biomass resources (including biofuels). Until recently, hydropower was the dominant renewable energy resource in the United States, but wind and solar energy are quickly overtaking it. The laws incentivizing, discouraging, and managing these renewable and non-renewable energy resources are discussed in this book.

It is also important to highlight that these different energy resources are used in different energy sectors. Historically, energy use could be divided fairly neatly into two main sectors—electricity and transportation. Coal, nuclear, and renewable energy resources are used almost exclusively in the electricity sector. By contrast, oil and biofuels are used almost exclusively in the transportation sector. Natural gas provides an exception to this rule of division as it is used increasingly in the electricity sector, is used to a growing extent in the transportation sector (for buses and some trucks and cars), and is also used heavily for home heating and for industrial purposes. But this division of energy resources between the primary energy sectors of electricity and transportation may be breaking down as we move into the future. There are major efforts to "electrify" the transportation sector to decrease our reliance on oil and increase our use of electric vehicles. To the extent this shift continues, we will see increasing volumes of energy resources that are currently almost exclusively used in the electricity system being recruited to also power our transportation system.

III. Federal and State Laws Governing Energy Systems and Industries

This book explores in detail the federal and state laws that govern the extraction, sale, transportation, use, and disposal of energy resources. These energy resources are used primarily in the two different energy systems that we just introduced—the electricity system and the transportation system. With regard to the laws governing these energy systems, the Federal Power Act regulates the generation and use of hydroelectric energy and well as the wholesale sale of electric energy in interstate commerce and the transmission of electric energy in interstate commerce. The Natural Gas Act governs the interstate sale of natural gas and interstate natural gas pipelines and facilities, including natural gas import and export facilities. Other federal laws governing the production, sale, or use of energy include, inter alia, the Atomic Energy Act; federal hardrock mining laws; laws governing the leasing of onshore coal, oil, and gas on federal lands; the Outer Continental Shelf Lands Act governing leasing of federal offshore oil, gas, and renewable resources; federal laws that provide tax credits or deductions associated with the production or use of energy resources including fossil fuels, renewable electricity, and biofuels; federal energy efficiency standards for appliances promulgated by the U.S. Department of Energy (DOE); and federal laws governing cyber security, the development of the "smart grid," and the reliability of the electric grid. Various "energy policy acts" such as the Energy Policy Act of 1992, the Energy Policy Act of 2005, and the Energy Independence and Security Act of 2007 contain a variety of provisions governing renewable electricity, mandates for the use of biofuels in gasoline, energy efficiency requirements for appliances, and transmission line siting provisions that are incorporated into many of the federal statutes discussed in this book.

Because the energy industry and the public's use of energy resources create significant environmental "externalities" in the form of air and water pollution and other environmental impacts, many of the nation's federal environmental laws target and regulate those externalities. For instance, the Clean Air Act regulates air emissions from electric power plants used in the electricity system as well as the vehicles and fuels used in the transportation system. Beyond the Clean Air Act, there are a vast number of environmental law that govern the energy industry and are discussed in more detail in

Chapter 4. These include, among others, the Clean Water Act; the Coastal Zone Management Act; the Endangered Species Act; and the Surface Mining Control and Reclamation Act. Moreover, the National Environmental Policy Act requires that any federal projects, federally-funded projects, or projects requiring any federal permit be subjected to a detailed evaluation of the potential adverse environmental impacts associated with the project prior to approval. The National Environmental Policy Act thus covers a significant number of energy-related decisions and projects.

Many aspects of energy production, sale, and use are regulated at the state and local levels instead of or in addition to the federal level. For instance, oil and gas exploration on non-federal lands is subject primarily to state property laws as well as state regulations governing spacing of wells, unitization (which involves combining various mineral tracts to ensure efficient development of oil and gas resources), safety, and environmental protection. State tort law in the form of nuisance, negligence, trespass, and waste also applies to oil and gas development. State and local zoning laws govern where and under what circumstances wind farms, solar plants, and fossil fuel plants can be built. States may impose clean energy mandates on electric utilities to require that a minimum percentage of electricity be generated from renewable energy resources, and states may also create property and income tax deductions and credits to incentivize the production or use of certain types of energy, in addition to any federal incentives or credits. Beyond energy production, state public utility commissions and other state energy agencies have authority to regulate retail sales of electricity and to approve or deny virtually all intrastate and interstate oil pipelines and electric transmission lines and intrastate natural gas pipelines.

IV. Federal and State Agencies Regulating the Energy Industry

Numerous federal and state regulatory agencies interpret, implement, and enforce the vast array of federal and state laws that govern the energy industry. One prominent federal agency in this area is the Federal Energy Regulatory Commission (FERC), which Congress created in 1930 as the Federal Power Commission (FPC). FERC has Congressional authority to regulate a wide variety of energy resources and industries, including wholesale electricity sales; interstate transmission of electricity, natural gas, and oil; the approval of interstate natural gas pipelines and liquefied natural gas (LNG) import and export facilities; the licensing of hydroelectric power plants; and the reliability of the electric grid.[1] FERC is an

---

[1] *What FERC Does*, FED. ENERGY REGULATORY COMM'N, http://www.ferc.gov/about/ferc-does.asp.

independent regulatory agency with five commissioners appointed by the President with the advice and consent of the Senate. Each commissioner serves a five-year term, and no more than three commissioners may belong to the same political party.[2]

Another federal agency that is central to the regulation of the energy industry is the DOE. The stated mission of the DOE is "to ensure America's security and prosperity by addressing its energy, environmental and nuclear challenges through transformative science and technology solutions."[3] DOE is responsible for funding and implementing a wide range of research and development projects both independently and in conjunction with state and local governments and private industry. These projects include, among others, developing advanced fuels, energy storage, advanced battery systems, alternative energy, energy efficiency, carbon capture and storage, and renewable energy. DOE is responsible for the National Laboratories—including Argonne, Fermi, Los Alamos, Oak Ridge, and Lawrence Berkeley National Laboratories—which house some of the nation's top scientists engaged in cutting-edge, energy-related research. DOE, often in connection with FERC, also plays a role in interstate electric transmission line siting and planning, as discussed in more detail in later chapters of this book.

Other federal agencies with statutory authority over energy matters include the Nuclear Regulatory Commission (NRC) (with authority over nuclear reactors and nuclear waste), the Department of Transportation (with authority over vehicle efficiency and rail and pipeline safety), and the U.S. Department of the Interior (with authority over the extraction and production of energy resources such as coal, natural gas, and oil located on federal public lands and located offshore in the Outer Continental Shelf). These agencies and their specific areas of authority are discussed throughout the book.

One additional federal agency deserves special note—the U.S. Environmental Protection Agency (EPA). Although EPA is best known for administering the bulk of the nation's federal environmental laws, it also plays an outsize role in the energy world. This is because the oil, gas, and electric industries emit significant amounts of air and water pollution in connection with the production, use, and disposal of the energy resources needed to produce electricity as well as to transport the nation's cars, trucks, buses, trains, water vessels, and airplanes.

At the state level, a variety of state energy agencies and public utility commissions administer a range of state energy laws

---

[2] *Commission Members*, FED. ENERGY REGULATORY COMM'N, http://www.ferc.gov/about/com-mem.asp.

[3] *Mission*, U.S. DEP'T OF ENERGY, http://www.energy.gov/mission.

governing retail electricity and natural gas sales; intrastate and interstate oil pipelines; intrastate natural gas pipelines; oil and natural gas production on state and private lands (and even federal lands, since states retain concurrent authority to regulate oil and gas operations on BLM lands); the siting of interstate and intrastate electric transmission lines; and the siting of energy generation facilities such as coal plants, natural gas plants, and wind and solar facilities.[4] State and local governments also are the primary regulators of oil and gas exploration on non-federal lands and in the coastal zone of each coastal state within 3 or 9 nautical miles of shore.[5]

The energy realm also contains regional and non-profit entities with regulatory authority, or, in the case of non-profit entities, with important management responsibilities for certain energy resources. In the electricity arena, regional transmission organizations (RTOs) and independent system operators (ISOs) are FERC-approved nonprofit entities that manage the regional transmission system for a wide range of electric utilities and other industry participants in approximately two-thirds of the country and run organized wholesale electricity markets within their territories. The North American Electric Reliability Corporation (NERC) is another FERC-approved non-profit entity and is responsible for promulgating and enforcing reliability standards for the nation's electric grid. Outside the electricity realm, the Delaware River Basin Commission and Susquehanna River Basin Commissions are the result of interstate compacts created in the 1960s by Maryland, New York, New Jersey, Delaware, and Pennsylvania to manage water resources in the river basin. In recent years these Commissions have been active in attempting to address water pollution and water use associated with increased natural gas production in the region resulting from hydraulic fracturing and horizontal drilling technologies.

V. Basic Administrative Law Concepts

All of the federal and state statutes and regulations discussed so far exist within a broader framework of administrative law. For readers not familiar with the field of administrative law, most federal energy laws begin with Congress enacting a statute setting forth one or more specific energy policies (in the area of electricity, transportation, nuclear energy, or the like) and granting regulatory authority to a federal agency with expertise in the field—such as FERC, DOE, or the Department of the Interior—to implement and

---

[4] States do not regulate nuclear power plants, which are approved by the NRC, or power plants on federal lands.

[5] Texas and Florida, for Florida's Gulf Coast, are the only states that have jurisdiction over waters from the coast and seaward 9 nautical miles.

enforce the law; promulgate regulations to provide more specific requirements, explanations, or incentives to further Congressional policy; and issue civil and criminal penalties for noncompliance with the statutory or regulatory provisions. For instance, this book will discuss various orders FERC has issued regarding open access to electric transmission lines, electric transmission line planning, and wholesale electricity sale ratemaking pursuant to its authority under the Federal Power Act. FERC has also issued rules governing the process applicants must follow if they are seeking a certificate of public convenience and necessity to build an interstate natural gas pipeline. EPA has issued regulations implementing the federal renewable fuel standard, which Congress created to mandate the use of biofuels in gasoline, and also to impose standards under the Clean Air Act for drilling certain types of oil and gas wells, among many other energy-related regulations.

In each instance, the agency is acting pursuant to its statutory authority in promulgating these rules and enforcing regulated parties' compliance with the rules. Not surprisingly, regulated parties, environmental groups, and others often argue that the agency does not have the authority it claims to have when it issues various rules and orders, or that the agency is implementing its rules in a manner that is not consistent with Congress's intent. These parties file lawsuits against the agency contesting its authority to take the action in question. With the exception of "independent" agencies such as FERC and the NRC, the vast majority of other federal agencies such as EPA and the Department of the Interior are under the control of the executive branch.[6] Thus, agency policy often changes when a new presidential administration begins, even though the agency continues to act pursuant to the same statute.

The overarching statute that governs agency rulemaking, adjudication, and judicial review when agency decisions are challenged is the Administrative Procedure Act.[7] Congress enacted the APA in 1946 at a time when Congress was creating a host of new regulatory agencies as part of the New Deal. The APA contemplates that agencies will engage in two primary activities—rulemaking and adjudication. The APA sets out the process agencies must follow in enacting rules and adjudicating disputes and also the standards courts should apply in reviewing agency decisions when challenged. Thus, federal agencies must comply with the statutory mandate Congress gives them in specific federal statutes as well as with the APA.

---

[6] *See* KRISTIN E. HICKMAN & RICHARD J. PIERCE, JR., FEDERAL ADMINISTRATIVE LAW 10–12 (2d ed. 2014) (discussing executive branch and independent agencies).

[7] 5 U.S.C. §§ 551, *et seq.* (West 2016).

The most common activity agencies engage in is rulemaking. When an agency engages in rulemaking it is exercising a legislative function pursuant to authority Congress granted to it in that area by statute. An agency generally acts using its rulemaking authority when it wishes to exercise a legislative function by creating policy of general applicability and future effect.[8] The APA defines a rule as "the whole or part of an agency statement of general or particular applicability and future effect designed to implement, interpret, or prescribe law or policy."[9] The APA provides for "formal rulemaking" where the rule is only issued after a trial-type hearing with evidence presented "on the record." It also provides for "informal" rulemaking (also referred to as "notice and comment rulemaking") where the rule is issued after notice is provided to the public, comments are received, and the rule is published along with responses to those comments and a statement of reasons why the rule is being issued and why the agency opted for the specific provisions in question.[10]

For several decades now, most federal agencies (including the ones discussed in this book) issue rules under the APA using informal rulemaking procedures. In many ways, the term "informal rulemaking" is a misnomer. Many of the rules issued by FERC, EPA, and other agencies require months or years to complete, involve responding to thousands (and sometimes hundreds of thousands) of public comments, and result in the agency preparing thousands of pages of explanation and justification for the final rule. In addition to the APA's rulemaking requirements, each agency has its own published procedural rules governing how it issues rules. Moreover, any rule that is defined as a "significant regulatory action" (i.e., that may have an annual effect on the U.S. economy of $100 million or more) is subject to a cost-benefit analysis by the Office of Information and Regulatory Affairs (OIRA) in the White House's Office of Management and Budget (OMB).[11] This often results in delays in issuing the rule as well as changes in the final rule.

Agencies engage in adjudication, in contrast to rulemaking, when a statute requires the agency to issue a decision, applicable in the first instance to a single party (or a small group of similarly situated parties), that is "on the record" after an opportunity for an

---

[8] *See* RICHARD J. PIERCE, JR., ADMINISTRATIVE LAW TREATISE § 6.1 at 402–03 (Wolters Kluwer 5th ed. 2010) (discussing agency rulemaking).

[9] 5 U.S.C. § 551(4) (West 2016).

[10] *See* 5 U.S.C. §§ 553(b), 556, and 557 (West 2016) (requirements for formal rulemaking); 5 U.S.C. §§ 553(b), 553(c) (West 2016) (requirements for informal rulemaking). *See also* HICKMAN & PIERCE, *supra* note 6, at 458–59, 465–67.

[11] *See* Exec. Order No. 12866, Regulatory Planning and Review, 58 Fed. Reg. 190 (Sept. 30, 1993); *About OIRA*, WHITE HOUSE OFFICE OF MGMT. & BUDGET, https://www.whitehouse.gov/omb/oira/about.

agency hearing.[12] The parties to an adjudication have an opportunity to provide testimony under oath, submit documentary evidence, and to subpoena and cross examine witnesses. Generally, an adjudication is retrospective in nature rather than prospective. An administrative law judge (ALJ) presides over the adjudication, and the rules of evidence apply, although they are construed more liberally than in a traditional courtroom proceeding. For instance, hearsay evidence, which would be excluded under the federal rules of evidence, is generally admissible in an agency adjudicatory proceeding. Depending on the specific agency's rules, the ALJ decision is either the final decision of the agency, or is a recommendation to the agency's decision-making body. In addition to the APA requirements, each agency has its own internal procedures governing how certain types of adjudications will proceed, and sometimes the statute itself requires certain procedures. For instance, the Federal Power Act has specific provisions governing the process for hearings involving the rates that FERC-regulated entities may charge their customers.

Regardless of whether the agency acts using its rulemaking or adjudicatory authority, parties aggrieved by the agency's action may challenge it in court under the APA. The judicial review provisions of the APA are significant because many of the nation's energy law statutes do not have a "private right of action" that allows the party to challenge an agency decision under the statute itself. In that situation, the aggrieved party may use the APA as a vehicle to file a lawsuit in federal court alleging that the agency exceeded its authority, acted in an unreasonable manner, or otherwise violated the law. This is true of the Federal Power Act, the Natural Gas Act, and many other natural resources and environmental law statutes enacted prior to the 1970s.

Under the APA, any person "suffering legal wrong because of agency action, or adversely affected or aggrieved by agency action within the meaning of a relevant statute is entitled to judicial review thereof."[13] It is important to note, however, that many statutes, including the Natural Gas Act, require a party to seek rehearing before the relevant agency or fulfill other procedural requirements before bringing an action in court. Moreover, there is a wealth of case law regarding which parties have constitutional "standing" to sue as a result of a particularized injury rather than a generalized grievance no different than any other U.S. citizen.[14] Many statutes, including the Federal Power Act and the Natural Gas Act, also contain requirements regarding where parties may challenge agency action.

---

[12] *See* 5 U.S.C. §§ 554, 556, 557 (West 2016); PIERCE, *supra* note 8, at 402–03.

[13] 5 U.S.C. § 702 (West 2016).

[14] *See, e.g.*, Sierra Club v. Morton, 405 U.S. 727 (1972); Lujan v. Defenders of Wildlife, 504 U.S. 555 (1992); Massachusetts v. EPA, 549 U.S. 497 (2007).

Sec. V   BASIC ADMINISTRATIVE LAW CONCEPTS   11

For instance, lawsuits challenging FERC orders under the Natural Gas Act or the Federal Power Act must be filed in either the U.S. Court of Appeals for the District of Columbia Circuit or in the U.S. Court of Appeals for the circuit in which the regulated entity is located or has its principal place of business. The party challenging agency action must also establish that the matter has not been "committed to agency discretion" although courts have interpreted that provision very narrowly. Thus, agency discretion generally is not a barrier to seeking judicial review so long as Congress has created some legal standard, even a broad one, to guide the agency.[15]

Once a party meets the requirements to challenge federal agency action in court, the APA provides that the court may compel agency action "unlawfully withheld or unreasonably delayed"[16] and also that it may set aside agency action if it is: (1) arbitrary, capricious, an abuse of discretion, or otherwise not in accordance with law; (2) contrary to a constitutional right, power, privilege or immunity; (3) in excess of statutory jurisdiction, authority, or limitations, or short of statutory right; (4) without observance of procedure required by law; (5) unsupported by substantial evidence in a formal rulemaking case or otherwise reviewed on the record of an agency hearing provided by statute; or (6) unwarranted by the facts to the extent that the facts are subject to trial de novo by the reviewing court.[17]

If a party is challenging a substantive rule by an agency rather than an agency's failure to follow proper procedure, the court's review of the agency decision is generally quite deferential as a result of the arbitrary, capricious, and abuse of discretion language in the APA. Nevertheless, the U.S. Supreme Court has held that courts must ensure that the agency took a "hard look" at the issue, based on the record, and explained why it resolved any contested issues as it did.[18]

Often, the dispute with the agency involves the agency's interpretation of language in the enabling statute that the agency contends supports the rule it issued. In that case, too, courts give a

---

[15] *See* Citizens to Preserve Overton Park, Inc. v. Volpe, 401 U.S. 410 (1971) (holding that the discretion exception to judicial review must be interpreted narrowly and does not bar review unless there is "no law to apply").

[16] 5 U.S.C. § 706(1) (West 2016). This provision generally applies when a federal statute sets a specific deadline for an agency to act and the agency misses that deadline. It generally does not apply simply because an agency declines to use its statutory authority to issue an order or decision a member of the public would like it to take. *See, e.g.,* Norton v. S. Utah Wilderness All., 542 U.S. 55 (2004) (where statute leaves the agency with discretion regarding when and how to act, environmental groups could not challenge failure of the agency to use its authority).

[17] 5 U.S.C. § 706(2) (West 2016).

[18] *See* Motor Vehicle Mfrs. Ass'n v. State Farm Mut. Auto Ins. Co., 463 U.S. 29 (1983); HICKMAN & PIERCE, *supra* note 6, at 560–62; PIERCE, *supra* note 8, §§ 7.1, 7.4 (discussing judicial review of agency rulemaking).

great deal of discretion to the agency's interpretation of the statutory language, since the agency has delegated authority to implement the statute by Congress and is considered to have technical expertise in that field. Under the standard the U.S. Supreme Court set forth in *Chevron U.S.A., Inc. v. NRDC*,[19] the court must first determine whether "Congress has directly spoken to the precise question at issue."[20] If Congress has, then the court must "give effect to the unambiguously expressed intent of Congress."[21] However, if the statute "is silent or ambiguous with respect to the specific issue," the question is then whether the agency's interpretation is "based on a permissible construction of the statute."[22] If it is, the court should uphold the agency's interpretation of the statutory language because the agency has delegated authority as a result of its expertise and must fill statutory gaps in a reasonable manner.[23] The Supreme Court and the lower courts have decided numerous cases elaborating on the *Chevron* doctrine and exploring the level of deference that should apply to agency actions in a variety of circumstances.[24]

Most states also have their own administrative procedure acts that apply to state agency action. Although certain state provisions may differ from the federal APA, they generally provide for agency rulemaking and adjudication and have similar provisions regarding judicial review of agency actions.

In the chapters that follow, readers will see a range of challenges to energy decisions made by agencies while also learning about the substantive aspects of energy law.

---

[19] 467 U.S. 837 (1984).

[20] *Chevron*, 467 U.S. at 842.

[21] *Id.* at 842–43.

[22] *Id.* at 843.

[23] *Id.* at 843–44. *See also* Pauley v. Bethenergy Mines, Inc., 501 U.S. 680 (1991) (applying *Chevron*).

[24] *See* PIERCE, *supra* note 8, §§ 3.3, 3.5 (discussing meaning of *Chevron*, scope of *Chevron*, and cases applying *Chevron*). *See also* KRISTIN E. HICKMAN & RICHARD J. PIERCE, JR., ADMINISTRATIVE LAW TREATISE §§ 3.5–3.6 (Wolters Kluwer 5th ed. 2015 Cum. Supp.) (discussing scope of *Chevron* and recent U.S. Supreme Court cases applying *Chevron*).

Chapter 1

ENERGY FACILITY SITING

I. Introduction

Energy facilities consist of the power plants used to generate energy, the pipelines and transmission lines used to transport that energy across or under land, and the import and export facilities used to facilitate the transport of energy across oceans.[1] Not surprisingly, energy facility siting is a mix of federal, state, and local regulation. The level of government that regulates the approval of energy facility siting depends in large part on the economics, politics, and history of the energy resource in question. As a result, siting authority is not consistent among different types of energy resources and transport modes. For instance, the federal government approves the siting of interstate natural gas pipelines while state and local governments approve the siting of interstate oil pipelines and interstate electric transmission lines. Although the federal government is heavily involved in the siting and regulation of nuclear facilities, state and local governments have exclusive siting authority over coal-fired power plants, wind and solar facilities, and oil and gas operations off federal lands. This Chapter explores the laws and regulations governing energy facility siting and highlights some of the interactions and conflicts that exist among the various governmental entities with authority over energy facility siting.

II. Energy Generation Facilities

The generation of electricity takes many forms, and different siting issues arise for different types of generation infrastructure. The bulk of U.S. generation takes place at centralized, "utility-scale" facilities that produce large quantities of electricity and use transmission lines to send electricity over long distances to population centers. A small yet growing amount of electricity generation is "distributed generation" (on-site, such as on rooftops or in parking lots), or "community-scale" generation, in which a medium-sized facility provides electricity to one or several neighborhoods or to a housing subdivision. Small- and medium-scale facilities are discussed in Chapter 8.

A developer of a utility-scale energy generation facility must consider numerous factors when deciding where to site the plant.

---

[1] Rail, barges, and trucks are also integral to the transportation of energy, but because these modes of energy transport are not "fixed" in the same manner as generation facilities, pipelines, and electric transmission lines, they are generally excluded from this Chapter's coverage of siting and permitting of energy facilities.

"Thermal" plants, which use a fuel such as natural gas, uranium, coal, or sunlight to heat up water to produce steam and turn a turbine, often must be sited near a water source such as a river or the ocean. Water is needed to produce steam and to cool the steam that turns the turbine. This cooling water is used to condense the steam before it is discharged into a lake, stream, the ocean, or other surface water source.[2] Other types of utility-scale generation, such as large-scale installations of solar photovoltaic panels or wind towers and turbines, do not require much water; water use for this type of generation infrastructure is typically limited to rinsing off dust and dirt that builds up on the generation equipment and reduces its efficiency.[3]

Developers of utility-scale generation consider many factors in addition to the presence of water at a site. They also must consider the presence and abundance of fuel and fuel delivery infrastructure. For example, a coal plant is most conveniently located near a railroad facility because coal is often shipped by rail,[4] and a natural gas plant needs access to a natural gas pipeline. Developers of these plants build a small "lateral" distribution line from the pipeline in order to access natural gas. Renewable energy plants need abundant fuel to be available on site. Wind and solar developers therefore rely on studies and maps that show the presence and abundance of wind or sunlight at different locations, and prior to building generation they install sophisticated scientific measurement stations on site to test the amount of sunlight or wind at a specific site.[5] Additionally, all developers of utility-scale electricity generation need a site with access to transmission lines to carry electricity from the site to other utilities, which then distribute retail electricity to customers.[6] Renewable energy developers often face the greatest transmission challenges because most abundant renewable electricity resources

---

[2] *Many Newer Power Plants Have Cooling Systems that Reuse Water*, U.S. ENERGY INFO. ADMIN. (Feb. 11, 2014), http://www.eia.gov/todayinenergy/detail.cfm?id=14971.

[3] INT'L ENERGY AGENCY, WATER FOR ENERGY: IS ENERGY BECOMING A THIRSTIER RESOURCE? 9 (excerpt from THE WORLD ENERGY OUTLOOK 2012), http://www.worldenergyoutlook.org/media/weowebsite/2012/WEO_2012_Water_Excerpt.pdf.

[4] *Rail Continues to Dominate Coal Shipments to the Power Sector*, U.S. ENERGY INFO. ADMIN. (Feb. 24, 2016), http://www.eia.gov/todayinenergy/detail.cfm?id=25092.

[5] *See, e.g.*, Ten Taxpayer Citizens Grp. v. Cape Wind Assocs., 373 F.3d 183, 186 (1st Cir. 2004) (describing a "scientific measurement device system" installed offshore to measure wind speed and other factors pending the proposed development of an offshore wind farm).

[6] For a discussion of the particular barriers that renewable energy developers face in terms of the need for access to transmission lines, *see* Alexandra B. Klass & Elizabeth J. Wilson, *Interstate Transmission Challenges for Renewable Energy: A Federalism Mismatch*, 65 VAND. L. REV. 1801 (2012).

tend to be located far from population centers and far from existing transmission lines.

Developers considering a site for utility-scale generation also must consider a variety of land use, environmental, and safety-based factors, such as whether electricity generation is an allowed use of land within a particular local zoning district (if local laws have not been preempted by state law),[7] or is off-limits due to state or federal land protections. Sometimes, a site is home to threatened and endangered species or is located within the flight path or nesting area of migratory birds or eagles. This requires the developer to work with federal and state authorities to reduce and/or prevent impacts to these species. In many cases, developers simply avoid sites if they know that certain protected species are located on a proposed site or near a proposed site. For example, wind developers contemplating building wind farms within a particular state tend to carefully survey large portions of the state for bat caves because wind turbines kill bats when bats fly too close to the turbines. These developers typically avoid portions of the state that have concentrated bat populations.

Once a power plant developer has identified potential sites based on available resources and likely obstacles to development on particular sites, the developer must obtain regulatory approval to site the power plant in the proposed location. As introduced above, states and local governments control the siting of most power plants, with the exception of nuclear energy. The federal government approves acceptable sites for nuclear plants and considers a variety of complex siting factors such as the proximity of the site to human populations; "seismic, meteorologic, hydrologic, and geologic data"; and potential barriers to emergency evacuation routes (such as drawbridges).[8]

States take three general approaches to approving particular sites for non-nuclear utility-scale development. Some states leave almost all decision-making to local governments. In many respects, this approach makes sense because local governments already have primary responsibility for regulating all types of land uses—commercial, industrial, and residential. States delegate land use planning and regulatory responsibilities to local governments, and

---

[7] If a local land use ordinance addresses commercial energy infrastructure and allows this infrastructure to be located within certain zoning districts, it typically allows the infrastructure as a permitted use (a use automatically allowed within that zone "as of right," provided the proper building permits and engineering approvals are issued), or as a conditional use, which must be approved within the zone on a case-by-case basis.

[8] NUCLEAR REGULATORY COMM'N, NUCLEAR POWER PLANT LICENSING PROCESS 6–7 (2004), http://www.nrc.gov/reading-rm/doc-collections/nuregs/brochures/br0298/br0298r2.pdf.

these governments establish zoning districts in which particular land uses are allowed or prohibited. Local governments then issue individual permits and approvals for construction and uses of land within these zoning districts. Therefore, in states like Oklahoma, Iowa, Texas, and Kansas, local governments primarily control the siting of certain forms of utility-scale generation—particularly renewable generation.[9] In Kansas, local governments are even allowed to ban certain types of generation, such as commercial-scale wind energy, from their county, town, or city.[10]

Some states take the opposite approach and preempt most local control over the siting of utility-scale electricity generation—or certain types of generation. For example, in Florida all thermal power plants that have a capacity to generate 75 or more megawatts of electricity, as well as solar power plants that have this capacity or a larger capacity, must be permitted by the state.[11] A developer must obtain from the state both a "certificate of need"—a certificate certifying that the plant is in fact "needed" from an economic perspective (e.g., that there is increasing demand for electricity)—and a siting approval. The state's Department of Environmental Protection (DEP) requires the developer of a proposed project subject to state siting requirements to submit a variety of documents on the proposed site. The Governor, along with a cabinet of three other executive officials who vote on certain important executive decisions in the state, must ultimately approve the site,[12] except in circumstances where an administrative law judge relinquishes jurisdiction to the state DEP.[13] Local governments participate in the administrative hearings leading up to the siting decision, and local governments have the opportunity to indicate that a proposed site for utility-scale infrastructure is inconsistent with their zoning.[14]

States that preempt local control over the siting of certain types of utility-scale infrastructure take a variety of approaches to determining the type of infrastructure that is subject to state siting control. Many, like Florida, preempt local control over the siting of large, thermal infrastructure (although allowing for local consistency determinations, as noted above).[15] Others, like Wyoming (which, like Florida, issues a state siting certificate for power plants but allows

---

[9] Hannah Wiseman et al., *Formulating a Law of Sustainable Energy: The Renewables Component*, 28 PACE ENVTL. L. REV. 827, 886 (2011).

[10] Zimmerman v. Bd. of County Comm'rs, 289 Kan. 926 (2009).

[11] FLA. STAT. § 403.506(1) (West 2016); F.S. § 403.503(14) (West 2016).

[12] FLA. STAT. § 403.509(2) (West 2016).

[13] FLA. STAT. § 403.509(1)(a) (West 2016).

[14] FLA. ADMIN. CODE ANN. r. 62–17.121(2) (West 2016).

[15] FLA. STAT. § 403.510(2) (West 2016).

some local influence over the siting decision[16]), apply state siting requirements to energy infrastructure that exceeds a threshold construction cost, which is also a proxy for the size of the project.[17] (More expensive projects tend to be larger and have more impacts, and thus tend to invoke the type of local opposition to siting that states wish to preempt and address through a centralized process.) Wyoming also requires wind farms with thirty or more wind turbines to receive an industrial siting certificate from the state.[18]

Under a third form of utility-scale power plant siting regulation, states take a "hybrid" approach, allowing local governments to have some control over certain aspects of siting but maintaining site control for other aspects. For example, in Wisconsin, local governments are preempted from regulating the distance by which wind turbines must be set back from homes and other infrastructure, and similar siting considerations, if those regulations are more stringent than the state regulations.[19] However, local governments in Wisconsin may sometimes promulgate more stringent siting restrictions than the state's restrictions if they can justify their regulations as being necessary to protect local health and safety and as meeting other requirements, such as not substantially increasing the cost of the system.[20] These governments also may deny the siting of a wind farm in an area designated for future commercial or residential use.[21]

In other hybrid states, power plant developers must comply with both local and state land use requirements, but they may use a streamlined centralized permitting process at which local concerns are considered within the state process. Some states, like Washington, allow developers of renewable power plants to choose to participate in a local approval process or a state permitting process that considers both state environmental, social, and other regulatory factors as well as local laws.[22] Through this process, local land use concerns are addressed, but the state's Energy Facility Site Evaluation Council may recommend preemption of local land use

---

[16] WYO. STAT. ANN. § 35–12–102(a)(vii)(E)(I) (West 2016).

[17] WYO. STAT. ANN. § 35–12–102 (West 2016) (defining an "industrial facility" that requires a state permit as "any industrial facility with an estimated construction cost of at least ninety-six million nine hundred thousand dollars ($96,900,000.00) as of May 30, 1987").

[18] WYO. STAT. ANN. § 35–12–102(a)(vii)(E)(1) (West 2016).

[19] WIS. STAT. § 66.0401 (West 2016).

[20] *Id.*

[21] WIS. ADMIN. CODE PSC § 218.33 (West 2016).

[22] *Energy Facility Site Evaluation Council: Siting/Review Process*, ACCESS WASHINGTON, http://www.efsec.wa.gov/cert.shtmlEnergy20Facility.

plans that are not consistent with the proposed renewable energy development.[23]

III. Electric Transmission Lines

The siting, approval, and eminent domain process for interstate electric transmission lines takes place almost exclusively at the state level. The Federal Power Act of 1935 grants FERC jurisdiction over "transmission of electric energy in interstate commerce" and "the sale of electric energy at wholesale in interstate commerce,"[24] but states retain authority over the location and construction of both intrastate and interstate electric transmission lines.[25]

In most states the legislature has granted authority to its public utility commission (PUC) or equivalent state agency to review and approve intrastate and interstate electric transmission lines within the state borders based on a variety of factors associated with showing a "need" for the line and the economics and environmental impacts associated with the line.[26] If the applicant meets the required conditions, the PUC issues a certificate[27] that allows the transmission line operator to construct the line and exercise eminent domain to acquire the property necessary to build the line if the operator is unable to enter into voluntary easements with all landowners in the line's path.[28] This certificate is typically called a "certificate of need" or "certificate of public convenience and necessity."[29] For lines that cross several states, the operator must seek certificates in multiple states using multiple standards.[30]

The law differs from state to state as to whether state PUCs should approve interstate transmission lines that have significant regional benefits, such as moving wind or solar power from resource rich parts of the country to populous areas where electricity demand

---

[23] *Id.*

[24] Federal Power Act § 201(b), 16 U.S.C. § 824(b)(1) (West 2016).

[25] The Federal Power Act expressly states that FERC "shall not have jurisdiction, except as specifically provided in this subchapter and subchapter III of this chapter, over facilities used for the generation of electric energy or over facilities used in local distribution or only for the transmission of electric energy in intrastate commerce, or over facilities for the transmission of electric energy consumed wholly by the transmitter." *Id.; see also* New York v. Fed. Energy Regulatory Comm'n, 535 U.S. 1, 5–8 (2002); Ill. Commerce Comm'n v. Fed. Energy Regulatory Comm'n, 721 F.3d 764, 773 (7th Cir. 2013).

[26] *See* Alexandra B. Klass & Jim Rossi, *Revitalizing Dormant Commerce Clause Review for Interstate Coordination*, 100 MINN. L. REV. 129, 130–31 (2015) (discussing state transmission line siting processes).

[27] *Id.* at 131.

[28] Alexandra B. Klass, *Takings and Transmission*, 91 N.C. L. REV. 1079, 1102 (2013); *see also id.* (discussing electric transmission siting laws and eminent domain authority in all 50 states).

[29] Klass & Rossi, *supra* note 26, at 131.

[30] *Id.* at 130–131.

is high (called "load centers") several states away, or whether PUCs should only find a "need" for the line when there are also significant in-state benefits.[31] Since the 1990s, both public utilities and "merchant" transmission line companies[32] have attempted to build multi-state, long distance transmission lines to improve grid reliability and to transport new sources of renewable energy, particularly wind, to load centers.[33]

Despite the uncertainties in state law, there have been several major projects in recent years that highlight the growing role of interstate transmission lines to provide regional reliability and renewable energy benefits. For instance, within the territory of the transmission line operator called the Midcontinent Independent System Operator (MISO),[34] an area covering most or part of over 10 states, public utilities and other industry stakeholders have worked with state regulators and regional planners to build a series of Multi-Value Project (MVP) lines designed to improve grid reliability and transport wind energy throughout the region.[35] These lines have taken more than a decade to plan and construct, and required approval by multiple state PUCs within the MISO region.[36] In Texas, the state has worked with merchant transmission line companies and public utilities to build its $6.8 billion Texas Competitive Renewable Energy Zone (CREZ) project over nearly a decade.[37] This project has spurred transmission utilities to construct 3,600 miles of

---

[31] See Klass, *supra* note 28, at 1107–1112.

[32] Merchant transmission companies generate revenue solely from contracts with electricity generators to transmit electricity over the transmission lines they build for ultimate delivery of electricity to the retail electricity market by other companies. Thus, unlike state-regulated public utilities, merchant transmission companies do not receive a regulated, cost-based rate of return from electricity users. *See* Heidi Werntz, *Let's Make a Deal: Negotiated Rates for Merchant Transmission*, 28 PACE ENVTL. L. REV. 421, 425 n.13 (2011).

[33] *See* Alexandra B. Klass, *The Electric Grid at a Crossroads: A Regional Approach to Siting Transmission Lines*, 48 U.C. DAVIS L. REV. 1895, 1925–28 (2015) (describing projects).

[34] See Chapters 3 and 6 for a discussion of RTOs and ISOs and their role in the U.S. electric grid. *See also Electric Power Markets: National Overview*, FED. ENERGY REGULATORY COMM'N, http://www.ferc.gov/market-oversight/mkt-electric/overview.asp; *Electric Power Markets: Midcontinent (MISO)*, FED. ENERGY REGULATORY COMM'N, www.ferc.gov/market-oversight/mkt-electric/midwest.asp.

[35] *See* Klass, *supra* note 33, at 1927 (providing a similar description); Ill. Commerce Comm'n v. FERC, 721 F.3d 764, 771–75 (7th Cir. 2013) (describing MISO's MVP projects and upholding FERC's approval of MISO's regional cost allocation to pay for those lines).

[36] *See Minnesota-Iowa Transmission Line*, CTR. FOR RURAL AFFAIRS, http://www.cfra.org/clean-energy-transmission-map/line/minnesota-iowa (discussing the siting and permitting process for one of several MISO MVP transmission lines); *Multi Value Project Portfolio Analysis*, MISO, https://www.misoenergy.org/Planning/TransmissionExpansionPlanning/Pages/MVPAnalysis.aspx (discussing eight-year planning process prior to permitting and construction).

[37] Klass, *supra* note 33, at 1926 (providing a similar description).

high-voltage transmission lines across Texas to connect approximately 16,000 MW of wind energy to the Texas grid.[38] And a merchant transmission line company, Clean Line Energy Partners, is attempting to build "five separate DC [direct current] high-voltage" transmission lines across multiple states in the Midwest and southeast "to bring wind energy to population centers,"[39] although it has faced opposition from landowners and state PUCs in many areas.

There are a few circumstances where the federal government has siting authority over electric transmission lines. As explained in previous work, first, developers of electric transmission lines that cross federal lands must obtain a siting permit from the federal agency with authority over the land in question,[40] often the Bureau of Land Management within the U.S. Department of the Interior.[41] Transmission line developers also may apply to FERC to obtain a license to build transmission lines that move electricity from hydroelectric facilities to the grid.[42]

Further, provisions of the Energy Policy Act of 2005 (EPAct 2005) grant the Western Area Power Administration and the Southwestern Power Administration, which are two federal agencies that sell electricity to utilities within their region, authority to "design, develop, construct, operate, maintain, or own . . . an electric power transmission facility and related facilities . . . needed to upgrade existing transmission facilities."[43] In March 2016, the Southwestern Power Administration, acting through DOE, exercised its authority under these provisions for the first time by approving Clean Line Energy Partners' Plains & Eastern Clean Line project proposed to transport wind power from Western Oklahoma to the

---

[38] *Id.*; *see also* Daniel Cusick, *New Power Lines Will Make Texas the World's 5th-Largest Wind Power Producer*, CLIMATEWIRE (Feb. 25, 2014), http://www.eenews.net/stories/1059995041; Matthew L. Wald, *Texas is Wired for Wind Power, and More Farms Plug In*, N.Y. TIMES (July 23, 2014), http://www.nytimes.com/2014/07/24/business/energy-environment/texas-is-wired-for-wind-power-and-more-farms-plug-in.html (reporting on completion of Texas CREZ projects); R. Ryan Staine, Note, *CREZ II, Coming Soon to a Windy Texas Plain Near You? Encouraging the Texas Renewable Energy Industry Through Transmission Investment*, 93 TEX. L. REV. 521 (2014) (discussing the CREZ process).

[39] Klass, *supra* note 33, at 1927 & n.165 (discussing Clean Line Energy Partner projects); *Projects*, CLEAN LINE ENERGY PARTNERS, http://www.cleanlineenergy.com/projects.

[40] Klass, *supra* note 33, at 1918.

[41] Klass & Wilson, *supra* note 6, at 1825–26.

[42] Klass, *supra* note 33, at 1918; 16 U.S.C. § 797(e) (West 2016); James J. Hoecker & Douglas W. Smith, *Regulatory Federalism and Development of Electric Transmission: A Brewing Storm?*, 35 ENERGY L.J. 71, 82 (2014).

[43] Energy Policy Act of 2005, 42 U.S.C. § 16421(a) (West 2016). These federal power marketing agencies also have statutory authority to participate with private parties in designing, constructing, or operating these projects. 42 U.S.C. § 16421(b) (West 2016).

Arkansas-Tennessee border.[44] In doing so, DOE overrode a decision by Arkansas regulators rejecting the project. DOE's decision will undoubtedly be subject to numerous legal challenges, but it illustrates a recent effort by DOE to use the limited siting authority it has to approve interstate electric transmission lines designed to transport renewable energy.

Finally, other provisions of EPAct 2005 grant DOE the authority to designate National Interest Electric Transmission Corridors (NIETCs) in parts of the country experiencing significant transmission congestion.[45] If DOE designates a NIETC, FERC then has authority to site transmission lines within the NIETC if states in the path of the transmission line fail to make a determination as to whether the line may be sited—i.e., by "withholding" approval.[46] Congress enacted these provisions of EPAct 2005 due to grid reliability concerns that arose after a large blackout in the Northeast in 2003, which left millions of customers without power.[47] Although DOE attempted to designate two NIETCs, and FERC attempted to create rules to govern the federal siting process, federal courts interpreted these federal siting provisions very narrowly, thus rendering the provisions a virtual nullity at present.[48]

Although there is no federal authority over the siting and eminent domain process for most electric transmission lines, FERC has issued a series of orders encouraging and then mandating utilities to participate in regional planning processes for interstate transmission lines. In particular, FERC Order 1000 promulgated in 2011 requires the following, as summarized in earlier work:

> Each public utility transmission provider must participate in a regional transmission planning process through an RTO or otherwise; establish transmission needs based on

---

[44] *Plains & Eastern Clean Line Transmission Line*, U.S. DEP'T OF ENERGY, http://energy.gov/oe/services/electricity-policy-coordination-and-implementation/transmission-planning/section-1222-0; *Plains & Eastern Clean Line*, CLEAN LINE ENERGY PARTNERS, http://www.plainsandeasterncleanline.com/site/home.

[45] Klass, *supra* note 33, at 1918; Energy Policy Act of 2005, 16 U.S.C. § 824p.

[46] Klass, *supra* note 33, at 1918–1919; Piedmont Envtl. Council v. Fed. Energy Regulatory Comm'n, 558 F.3d 304, 314 (4th Cir. 2009), *cert. denied*, 558 U.S. 1147 (2010).

[47] Klass, *supra* note 33, at 1918; U.S.-CANADA POWER SYSTEM OUTAGE TASK FORCE, FINAL REPORT ON THE AUGUST 14, 2003 BLACKOUT IN THE UNITED STATES AND CANADA: CAUSES AND RECOMMENDATIONS 1 (Apr. 2004), http://energy.gov/sites/prod/files/oeprod/DocumentsandMedia/BlackoutFinal-Web.pdf.

[48] Klass, *supra* note 33, at 1919; Cal. Wilderness Coal. v. U.S. Dep't of Energy, 631 F.3d 1072 (9th Cir. 2011) (invalidating DOE's NIETCs in the Southwest and Mid-Atlantic regions after finding that DOE did not meet state consultation requirements); Piedmont Envtl. Council v. Fed. Energy Regulatory Comm'n, 558 F.3d 304, 314 (4th Cir. 2009) (invalidating FERC rule allowing FERC to approve electric transmission lines in NIETCs where a state has denied (as opposed to simply failed to act upon) a siting permit).

"public policy" requirements, which include state RPSs and other federal laws and regulations; and coordinate with transmission providers in neighboring regions to determine the most cost-effective solutions to mutual transmission needs. Order 1000 also eliminates certain utilities' "right of first refusal" to build transmission lines within their territories in order to allow market forces to spur the development of new transmission lines.[49]

Although regional plans developed under Order 1000 might encourage states to site new transmission lines, regional organizations still cannot require states to approve the siting of new transmission lines, and thus barriers to siting remain.

IV. Oil and Gas Pipelines

The legal regime for the siting of oil pipelines is similar to the transmission line regime because state and local governments retain authority over siting. Interstate natural gas pipelines, in contrast, are subject to wholly federal siting authority.

A. Oil Pipelines

Pipeline operators that wish to construct intrastate or interstate oil pipelines must follow state law, where it exists, that governs the application and approval processes to build these pipelines. Different states require different types of notices, permits, and approvals for pipeline construction and operation, which means an interstate oil pipeline must obtain all necessary approvals, siting permits, and rights of way from multiple state and local governments.[50] States vary significantly in their approval processes. For instance, Texas does not require a siting permit or other approval prior to constructing an oil pipeline.[51] The company must only notify the Texas Railroad Commission of any oil pipeline construction project longer than one mile[52] and must report whether it is a "gas utility, common carrier or private line."[53] By contrast, Illinois and Minnesota are examples of states that require a pipeline operator to seek a certificate of public convenience and necessity or a certificate of need

---

[49] Klass, *supra* note 33, at 1938–39; Transmission Planning and Cost Allocation by Transmission Owning and Operating Utilities, Order No. 1000, 136 FERC 61,051 (2011).

[50] For a summary of the oil pipeline siting and eminent domain laws in all 50 states, see Alexandra B. Klass & Danielle Meinhardt, *Transporting Oil and Gas: U.S. Infrastructure Challenges*, 100 IOWA L. REV. 947, App. (2015).

[51] *Id.* at 984.

[52] *Id.* (citing *Pipeline Eminent Domain and Condemnation*, RAILROAD COMM'N OF TEX., http://www.rrc.state.tx.us/about-us/resource-center/faqs/pipeline-safety-faqs/faq-pipeline-eminent-domain-and-condemnation/).

[53] RAILROAD COMM'N OF TEX., *supra* note 52.

from a state commission that reviews the potential economic and environmental impacts of the proposed oil pipeline, often after a trial-type proceeding before an administrative judge who then makes a recommendation to the commission.[54]

The issue of eminent domain arises frequently in the context of building new oil pipelines. In most states (with the exception of Colorado, as discussed below), oil pipelines have the right to exercise eminent domain to obtain the easements necessary to build intrastate and interstate pipelines because they are defined by state statute as a "public use," even though it is a private entity (the pipeline company) rather than a governmental entity that is seeking to obtain the property.[55] In many states, but not all, the pipeline company must obtain a certificate of need or a siting permit prior to exercising eminent domain authority.[56] While most pipeline companies are able to negotiate voluntary easement agreements with landowners in the path of the oil pipeline, eminent domain is an important tool that creates a disincentive for landowners to "hold out" and attempt to recover easement compensation amounts significantly greater than the fair market value of the easement or to stop the pipeline completely.[57]

Notably, in 2012, the Colorado Supreme Court interpreted its pipeline siting law in a manner that did not allow oil pipelines to exercise eminent domain authority. In *Larson v. Sinclair Transportation Co.*,[58] an oil pipeline company sought to exercise eminent domain to build a new oil pipeline under the landowner's property parallel to a pipeline that already existed under the property pursuant to an easement.[59] The landowner challenged the use of eminent domain for the new easement under the relevant statute, which provided:

> Any foreign or domestic corporation organized or chartered for the purpose, among other things, of conducting or maintaining a pipeline for the transmission of power, water, air, or gas for hire to any mine or mining claim or for any manufacturing, milling, mining, or public purpose shall have the right-of-way for the construction, operation, and maintenance of such pipeline for such purpose through any

---

[54] Klass & Meinhardt, *supra* note 50, at 987, 1037; *see also id.* at App. (detailing laws governing pipeline siting and eminent domain in additional states).

[55] Klass & Meinhardt, *supra* note 50, at 983.

[56] *Id.*

[57] *Id.* Under the U.S. Constitution and most state constitutions, private property may only be taken for a "public use" and upon payment of just compensation. *See* U.S. CONST. amend. V.

[58] 284 P.3d 42 (Colo. 2012) (en banc).

[59] *See* Klass & Meinhardt, *supra* note 50, at 987 (describing the case).

lands without the consent of the owner thereof, if such right-of-way is necessary for the purpose for which said pipeline is used.[60]

The court found that the Colorado General Assembly's grant of eminent domain authority only "intended to authorize eminent domain power for the construction of electric power infrastructure," despite its reference to "pipelines," determining that "pipelines" referred to "pipes involved in delivering electric power through the power grid, such as those pipes encasing underground electric wiring" and not oil pipelines.[61] Although Colorado legislators introduced a bill to expressly grant eminent domain authority to oil pipelines in the next legislative session, it failed to pass.[62] As a result, Colorado law is now clear that oil pipeline companies lack eminent domain authority.[63] As noted earlier, most state laws expressly grant eminent domain authority to oil pipelines, but in some states, the language of the relevant statutes is more ambiguous, as was the case in Colorado, and may be ripe for challenges in the future.

These types of challenges to eminent domain authority by oil pipelines are growing as a result of the increasing opposition to new oil pipelines and other energy transport infrastructure in many communities. Much of the oil and gas pipeline infrastructure in the United States was constructed in the middle of the 20th century, prior to the environmental movement of the 1970s and the existence of numerous, well-organized environmental groups. However, the widespread use of hydraulic fracturing technologies beginning in the late 2000s, discussed in more detail in Chapter 2, has led to a significant increase in oil and gas production in the United States as well as efforts to significantly expand the pipeline networks needed to transport these new sources of energy. Because oil pipelines are sited and approved at the state level, it allows landowners and environmental groups a greater opportunity to participate in the regulatory process and oppose these pipelines than exists under the federal approval process that governs interstate natural gas pipelines.

A prime example of the more widespread opposition to oil pipelines by landowners and environmental groups and the impact

---

[60] COLO. REV. STAT. § 38–4–102 (2013). A related provision grants eminent domain authority to "pipeline compan[ies]." *Id.* § 38–5–105.

[61] 284 P.3d at 45.

[62] Tessa Cheek, *Senate Caucuses on Budget; House Approves Easier Cross-State Firearms Purchases and Higher Ed Funding, Kill's Pipeline Bill*, THE COLO. INDEP. (Apr. 2, 2014), http://www.coloradoindependent.com/146826/senate-caucuses-on-budget-house-approves-easier-cross-state-firearms-purchases-and-higher-ed-funding-kills-pipeline-bill (noting that Colorado House "effectively killed the bill by moving to delay a vote until May 9, two days after the session ends").

[63] Klass & Meinhardt, *supra* note 50, at 987.

of that opposition on the regulatory process is the Keystone XL pipeline.[64] The developer of the project, TransCanada Corporation, began seeking permits and approvals in 2008.[65] Keystone XL is an international pipeline proposed by TransCanada Corporation to run 1,179 miles from Hardesty, Alberta, to Steele City, Nebraska where it would meet existing TransCanada pipelines that transport crude oil to Oklahoma and ultimately to Gulf Coast refineries.[66] The pipeline would carry crude oil from the Canadian oil sands (or "tar sands") region in Alberta as well as smaller quantities of shale oil produced in the Williston Basin in North Dakota.[67] Keystone XL has a different approval process than many other U.S. interstate oil pipelines because it crosses an international border and thus must obtain a U.S. State Department Presidential Permit[68] in addition to the multiple state regulatory approvals required. Whether the State Department issues a Presidential Permit is based on whether the project would serve the "national interest,"[69] which is subject to executive branch interpretation. In deciding whether to issue a Presidential Permit, the State Department completed an Environmental Impact Statement (EIS) under the National Environmental Policy Act to analyze the cumulative environmental effects of the entire span of the pipeline project rather than simply the impacts of the facilities located at the border.[70] When significant opposition by environmental groups and landowners in Nebraska and at the federal level slowed review of the project, TransCanada split the project into the 1,179 mile pipeline project from Alberta to Nebraska and two additional projects further south from Nebraska to the Gulf Coast (the Houston Lateral Project and the Gulf Coast Project).[71] The latter two projects required only state approvals; the Gulf Coast Project has since been built and is now operating, and the Houston Lateral is expected to be completed in 2016.[72]

---

[64] For a similar, longer summary of the project and the associated legal process, *see* Klass and Meinhardt, *supra* note 50, at 984–986.

[65] Exec. Order No. 13337, 69 Fed. Reg. 25299 (Apr. 30, 2004).

[66] *Keystone XL Pipeline Project*, TRANSCANADA, http://www.transcanada.com/keystone.html.

[67] U.S. Dep't of State, Final Supplemental Environmental Impact Statement for the Keystone XL Project, Executive Summary ES-6 to ES-7 (Jan. 2014), https://keystonepipeline-xl.state.gov/documents/organization/221135.pdf.

[68] Exec. Order 13337, *supra* note 65.

[69] Exec. Order 13337(g), *supra* note 65.

[70] U.S. Dep't of State, *supra* note 67, at ES-6 to ES-7.

[71] Keystone Pipeline, L.P., Application of TransCanada Pipeline, L.P. for a Presidential Permit Authorizing the Construction, Connection, Operation, and Maintenance of Pipeline Facilities for the Importation of Crude Oil to be Located at the United States-Canada Border (May 4, 2012), http://www.keystone-xl.com/wp-content/uploads/2012/11/kxl-pp-application.pdf.

[72] *Houston Lateral and Terminal*, TRANSCANADA, http://www.transcanada.com/houston-lateral.html.

The State Department completed an EIS for the new project in January 2014,[73] but did not issue a decision until November 2015, when it denied the permit on grounds that the pipeline would not serve the national interest and would interfere with the President's ability to be a world leader in addressing climate change.[74] During the years TransCanada awaited a State Department decision on the initial and the revised permit application, lawsuits in Nebraska held up the state permitting process. Prior to Keystone XL, Nebraska was a state like Texas (described above) that did not have an approval process for oil pipelines. However, in 2011, the Nebraska Governor called a special session of the state legislature to enact the state's first pipeline siting legislation—the Major Oil Pipeline Siting Act—which requires pipelines proposed after November 2011 to file an application with the Nebraska Public Service Commission (an independent state agency) and receive approval, prior to construction, based on a review of the project.[75]

Then in 2012, the Governor signed into law a new pipeline bill, which amended the Major Oil Pipeline Siting Act to grant authority to the Nebraska Department of Environmental Quality and the Governor to approve pipelines.[76] The legislature also revised Nebraska's eminent domain law to require any "major" oil pipeline to receive approval from the Governor or the Public Service Commission prior to exercising eminent domain. Based on the 2012 law, the Nebraska Governor approved Keystone XL through Nebraska,[77] but landowners challenged the 2012 law on grounds that allowing the Governor to exercise authority over pipelines instead of the independently elected public service commission violated the state's constitution.[78] While litigation was pending over the validity of the revised siting law, TransCanada filed a request for approval from the public service commission to avoid controversy over the constitutionality of the new siting law.[79] The State Department's rejection of the Presidential Permit put those state proceedings on

---

[73] U.S. Dep't of State, *supra* note 67.

[74] *Record of Decision and National Interest Determination*, U.S. DEP'T OF STATE, https://keystonepipeline-xl.state.gov/nid/249254.htm.

[75] L.B. 1, codified at NEB. REV. STAT. § 57–1405 (Supp. 2012).

[76] L.B. 7, codified at NEB. REV. STAT. § 57–1603 (Supp. 2012).

[77] Letter from Governor Heineman to President Barack Obama and Secretary Hillary Rodham Clinton (Jan. 22, 2013), http://www.keystone-xl.com/wp-content/uploads/2012/11/Governor_Pipeline_Approval.pdf.

[78] Second Amended Complaint for Declaratory Judgment at P13, Thompson v. Heineman, No. CI 12–2060 (Neb. Dist. Ct. Mar. 18, 2013), 2013 WL 7230784.

[79] *TransCanada Will Apply to Nebraska Public Service Commission to Approve Keystone XL Route*, TRANSCANADA (Sept. 30, 2015), http://www.transcanada.com/announcements-article.html?id=1988664&t=.

hold,[80] but if a future president were to reverse the permit denial decision, the state law permitting issues would then come back into play.

In sum, Keystone XL provides an excellent example of the interplay between federal and state siting law for an international, multi-state oil pipeline. Moreover, as a result of the national publicity over Keystone XL, landowners and environmental groups have created new alliances to mount challenges to other oil pipeline projects across the country, and the spotlight on these types of projects may well lead to new pipeline laws in other states, as it has in Nebraska.

B. Natural Gas Pipelines

Unlike the siting process for interstate oil pipelines, one federal agency—FERC—controls the siting and approval process for interstate natural gas pipelines. Federal control over the siting of interstate natural gas pipelines dates back to the Natural Gas Act of 1938, when Congress first gave the Federal Power Commission (now FERC) authority to regulate sales of natural gas in interstate commerce, transportation of natural gas in interstate commerce, and the facilities used for sales and transportation.[81] Section 7 of the Natural Gas Act created the process by which a pipeline operator may seek a certificate of public convenience and necessity from FERC to build an interstate natural gas pipeline after a review of the economic and environmental impacts of the pipeline.[82] Amendments to the Natural Gas Act in 1947 allowed FERC to grant nationwide eminent domain authority to pipelines receiving a certificate of public convenience and necessity, which allowed pipelines to overcome state opposition to natural gas transport infrastructure that had led to gas shortages on the east coast and consequential industry shutdowns in the 1940s.[83]

---

[80] *Gov. Ricketts Comments on TransCanada's Decision to Withdraw Keystone XL Application*, OFF. OF GOV. PETE RICKETTS (Nov. 18, 2015), https://governor.nebraska.gov/press/gov-ricketts-comments-transcanada%E2%80%99s-decision-withdraw-keystone-xl-application.

[81] *See, e.g.,* 15 U.S.C. § 717c (West 2016) (providing federal authority over natural gas rates and charges); 15 U.S.C. §§ 717f(c)–(h) (West 2016) (providing federal authority over natural gas facilities); Minisink Residents for Envtl. Pres. & Safety v. Fed. Energy Regulatory Comm'n, 762 F.3d 97, 101–02 (D.C. Cir. 2014) (describing federal process for siting and approving interstate natural gas pipelines under the Natural Gas Act).

[82] 15 U.S.C. § 717f (West 2016); *supra* note 81.

[83] 15 U.S.C. § 717f(h) (West 2016); Klass, *supra* note 33, at 1906–07 (describing landowner and state opposition to interstate pipelines in the 1940s and subsequent amendments to the Natural Gas Act in 1947 to create federal eminent domain authority for such pipelines).

As of 2007, over 80 percent of the natural gas in the United States travels by pipelines to its final destination.[84] There are three stages to the natural gas pipeline permitting process: (1) pre-filing; (2) application; and (3) post-authorization. As the Government Accountability Office explains, "In 2002, FERC created a voluntary pre-filing phase in its permitting process to 'facilitate and expedite the review of natural gas pipeline projects through early coordination with FERC and cooperating agencies.'"[85] The rationale behind the pre-filing process is to allow stakeholders (e.g., the applicant, environmental groups, citizens, states, other federal agencies) to be involved in the certificate process as early as possible and collect as much information in advance in order to shorten the overall timeline for review and decision. FERC must approve a company's pre-filing request, and pipeline companies that wish to use it must initiate it seven to eight months before filing a certificate application.

In deciding whether to grant or deny an application for a certificate, FERC considers "the project's potential impact on pipeline competition, the possibility of overbuilding, subsidization by existing customers, potential environmental impacts, avoiding the unnecessary use of eminent domain, and other considerations."[86] If FERC issues a certificate, the pipeline must file an implementation plan with FERC that addresses any required environmental mitigation and must also file weekly reports on inspection and compliance during the construction process.[87]

V. LNG Terminals

LNG is natural gas that has been cooled to at or below -260° F to turn it into a liquid state, significantly reduce its volume, and allow it to be stored, transported in specialized tankers across oceans, and then re-gasified and transported once again by pipeline after it

---

[84] *Interstate Natural Gas Pipeline Segment, Natural Gas*, U.S. ENERGY INFO ADMIN., https://www.eia.gov/pub/oil_gas/natural_gas/analysis_publications/ngpipeline/interstate.html.

[85] U.S. GOV'T ACCOUNTABILITY OFF., GAO–13–221, PIPELINE PERMITTING: INTERSTATE AND INTRASTATE NATURAL GAS PERMITTING PROCESSES INCLUDE MULTIPLE STEPS, AND TIME FRAMES VARY 12 (2013), http://www.gao.gov/assets/660/652225.pdf. *See also* 18 C.F.R. § 157.21 (2014) (describing pre-filing procedures).

[86] PAUL W. PARFOMAK, CONGR. RESEARCH SERV., INTERSTATE NATURAL GAS PIPELINES: PROCESS AND TIMING OF FERC PERMIT APPLICATION REVIEW 6 (Jan. 16, 2015), https://www.fas.org/sgp/crs/misc/R43138.pdf (internal quotations and citations omitted); *see also id.* at 2–5 (detailing FERC procedures and criteria for natural gas pipeline certificates); CAROLYN ELEFANT, KNOWING AND PROTECTING YOUR RIGHTS WHEN AN INTERSTATE GAS PIPELINE COMES TO YOUR COMMUNITY (May 17, 2010), http://lawofficesofcarolynelefant.com/wp-content/uploads/2010/06/FINALTAGguide.pdf (discussing pipeline permitting, construction, and eminent domain process).

[87] PARFOMAK, *supra* note 86, at 6.

reaches its destination.[88] There are two types of LNG terminals. The first type, which converts natural gas into LNG, is an export facility, typically called a "liquefaction terminal." The second type, which handles imports and converts LNG back into natural gas, is called a "regasification terminal."[89]

As discussed in more detail in Chapter 7, the U.S. Department of Energy (DOE) approves the import and export of natural gas through its Office of Fossil Energy, and the Natural Gas Act grants FERC authority to approve any LNG terminal used to import or export natural gas.[90] If the United States has a Free Trade Agreement (FTA) with the foreign nation for natural gas the application for import or export is automatically deemed consistent with the "public interest" and DOE must grant it without delay.[91] Exports to non-FTA nations are presumed to be in the public interest unless, after opportunity for a hearing, the DOE finds that the authorization would not be consistent with the public interest.[92]

Under Section 3(e) of the Natural Gas Act, FERC has "exclusive authority to approve or deny an application for the siting, construction, expansion, or operation of an LNG terminal."[93] This section of the Natural Gas Act was added as part of EPAct 2005, as Congress attempted to clarify disputes between FERC and the states over the authority for LNG terminal siting. These amendments also included a definition of "LNG terminal" in Section 2 of the Natural Gas Act, which states that an LNG terminal includes "all natural gas facilities located onshore or in State waters that are used to receive, unload, load, store, transport, gasify, liquefy, or process natural gas that is imported to the United States from a foreign country, [or]

---

[88] FEDERAL ENERGY REGULATORY COMM'N, A GUIDE TO LNG: WHAT ALL CITIZENS SHOULD KNOW 1 (Apr. 29, 2005), http://www.mudomaha.com/sites/default/files/lng.pdf; *see also* Rachel Clingman & Audrey Cumming, *The 2005 Energy Policy Act: Analysis of the Jurisdictional Basis for Federal Siting of LNG Facilities*, 2 TEX. J. OIL GAS & ENERGY LAW 58, 60 (2007).

[89] DIV. OF ENERGY MKT. OVERSIGHT, FED. ENERGY REGULATORY COMM'N, ENERGY PRIMER: A HANDBOOK OF ENERGY MARKET BASICS 18 (2012), http://www.ferc.gov/market-oversight/guide/energy-primer.pdf.

[90] *See, e.g.*, Sabine Pass Liquefaction, DOE/FE Order No. 3669, (U.S. Dep't of Energy June 26, 2015) (order granting approval of LNG exports and explaining in detail the DOE and FERC approval processes).

[91] 15 U.S.C. § 717b(c) (West 2016). DOE regulations implementing the natural gas import and export provisions of the Natural Gas Act are at 10 C.F.R. pt. 590. Free Trade Agreement countries for natural gas are Australia, Bahrain, Canada, Chile, Dominican Republic, El Salvador, Guatemala, Honduras, Jordan, Mexico, Morocco, Nicaragua, Oman, Peru, Singapore, and South Korea. *See How to Obtain Authorization to Import and/or Export Natural Gas and LNG*, U.S. DEP'T OF ENERGY, http://energy.gov/fe/services/natural-gas-regulation/how-obtain-authorization-import-andor-export-natural-gas-and-lng.

[92] 15 U.S.C. § 717b(a) (West 2016).

[93] 15 U.S.C. § 717b(e)(1) (West 2016).

exported to a foreign country from the United States."[94] Section 2 also states that an LNG terminal does not include any pipeline or storage facility subject to FERC jurisdiction under Section 7 of the Natural Gas Act.[95]

Section 3(e) of the Natural Gas Act directs FERC to act on applications for LNG terminals after a hearing and on terms and conditions FERC finds necessary and appropriate.[96] FERC regulations implementing Section 3(e) require the applicant to include in its application a statement demonstrating that the proposal "is not inconsistent with the public interest" and demonstrate that the proposal will "improve access to supplies of natural gas, serve new market demand, enhance the reliability, security, and/or flexibility of the applicant's pipeline system, improve the dependability of international energy trade, or enhance competition within the United States for natural gas transportation or supply."[97] The applicant also must demonstrate that the new terminal will not impair the applicant's ability to render transportation service in the United States to existing customers and that the facility will not restrict or prevent other United States companies from extending their activities in the same general area.[98]

Ever since the widespread use of hydraulic fracturing and directional drilling technologies in the late 2000s created massive new sources of U.S. natural gas, there has been a significant push by the gas industry to build new export terminals to sell natural gas abroad. Just prior to this boom, however, companies had begun to substantially expand gas imports. FERC had approved multiple new LNG import terminals—five new terminals "in the latter half of the 2000s"[99] and others that were "re-commissioned and expanded" during that time period.[100] Many existing import terminals have

---

[94] 15 U.S.C. § 717a(11) (West 2016). By contrast, LNG facilities located beyond state waters are licensed under the Deepwater Port Act, which requires the applicant to obtain a license from the United States Maritime Administration (MARAD) within the U.S. Department of Transportation, in conjunction with the U.S. Coast Guard, and also obtain approval from the governor or governors of the adjacent states. *See* 33 U.S.C. § 1501, *et seq.* (West 2016); MICHAEL RATNER, ET AL., CONGR. RESEARCH SERV., LNG EXPORTS PERMITTING PROCESS 3–5 (June 20, 2013), http://www.energy.senate.gov/public/index.cfm/files/serve?File_id=fb60c4c3-bff2-4fd5-b669-bf0049c4689b (explaining onshore and offshore LNG permitting processes).

[95] 15 U.S.C. § 717a(11)(B) (West 2016).

[96] 15 U.S.C. § 717b(e) (West 2016).

[97] 18 C.F.R. § 153.7(c)(1)(i) (West 2016).

[98] 18 C.F.R. § 153.7(c)(1)(ii), (iii) (West 2016).

[99] MICHAEL RATNER ET AL., CONG. RESEARCH SERV., U.S. NATURAL GAS EXPORTS: NEW OPPORTUNITIES, UNCERTAIN OUTCOMES, 2 (Jan. 28, 2015), https://www.fas.org/sgp/crs/misc/R42074.pdf.

[100] *Id.*; *see also* U.S. DEP'T OF ENERGY, QUADRENNIAL ENERGY REVIEW, APPENDIX B: NATURAL GAS INFRASTRUCTURE, at 23 and Table B–6 (2015), http://energy.gov/sites/prod/files/2015/09/f26/QER_AppendixB_NaturalGas.pdf.

since applied to convert to export facilities to align with the new influx of domestic natural gas.[101] Additionally, between the end of 2012 and 2016 FERC approved six new LNG export terminals in response to producer requests (and an additional five terminals that were not yet under construction as of 2016).[102]

Environmental groups and local residents have challenged the approval or expansion of LNG terminals in recent years, but none of those challenges have succeeded to date. For instance, the Sierra Club has challenged multiple LNG terminals including the Dominion Cove Point LNG terminal in Maryland, the Freeport LNG terminal in Texas, and the Sabine Pass and Cameron LNG terminals in Louisiana.[103] The Sierra Club has argued that increases in LNG exports will increase hydraulic fracturing of U.S. shale formations and cause adverse environmental impacts, including air pollution, water pollution, and climate change impacts.[104] DOE has concluded that it is not required to consider "upstream" impacts in its NEPA review of natural gas export requests—indeed, it believes that it cannot accurately estimate the exact amount of new wells that will be drilled and fractured as a result of the approval of additional exports, or the environmental impacts of those wells—but it nevertheless conducted a general evaluation of such upstream impacts in 2004.[105] In 2016, EPA encouraged FERC to measure and evaluate the climate impacts of LNG export terminals as part of FERC's revisions of its environmental review process for these

---

[101] RATNER ET AL., *supra* note 99, at 3, 7.

[102] *Energy Department Conditionally Authorizes Cameron LNG to Export Liquefied Natural Gas*, U.S. DEP'T OF ENERGY (Feb. 11, 2014), http://energy.gov/articles/energy-department-conditionally-authorizes-cameron-lng-export-liquefied-natural-gas; *North American LNG Import/Export Terminals Approved*, FED. ENERGY REGULATORY COMM'N, OFF. OF ENERGY PROJECTS (July 11, 2016), http://www.ferc.gov/industries/gas/indus-act/lng/lng-approved.pdf.

[103] *See, e.g.,* In re Sabine Pass Liquefaction, LLC, 139 FERC P 61039 (F.E.R.C.), 2012 WL 1312891 (Apr. 16, 2012) (FERC order granting authorization to construct LNG export facility); In re Sabine Pass Liquefaction Expansion, LLC 151 FERC P 61253 (F.E.R.C.), 2015 WL 3877112 (June 23, 2015) (order denying Sierra Club's request for rehearing on order approving LNG export facility expansion); *Stop LNG Exports*, SIERRA CLUB, http://content.sierraclub.org/naturalgas/stop-lng-exports (discussing Sierra Club's continued opposition to every proposed new LNG facility in the United States because of the adverse environmental effects of natural gas development on air, water, and climate); *Sierra Club Continues Fight Against LNG Exports: Files Protest Against Golden Pass LNG Export Project*, SUTHERLAND LNG (Aug. 12, 2014), http://www.lnglawblog.com/2014/08/sierra-club-continues-fight-against-lng-exports-files-protest-against-golden-pass-lng-export-project/.

[104] Mark Westlund, *Groups Appeal Federal Approval of Cove Point LNG Export Facility*, SIERRA CLUB (May 7, 2015), http://content.sierraclub.org/press-releases/2015/05/groups-appeal-federal-approval-cove-point-lng-export-facility; *In re Sabine Pass Liquefaction Expansion LLC*, *supra* note 103.

[105] Draft Addendum to Environmental Review Documents Concerning Exports of Natural Gas from the United States for Public Comment, 79 Fed. Reg. 107 at 32258 (June 4, 2014).

projects.[106] To date, FERC has declined to conduct such analysis in its NEPA review of LNG export terminals and, in 2016, the D.C. Circuit upheld that decision in response to legal challenges associated with facilities in Maryland and Texas.[107]

VI. Coal Transport: Rail and Coal Export Facilities

The relatively streamlined federal process for approving natural gas pipelines and LNG terminals stands in contrast to the process for approving coal export terminals, which requires approvals from federal, state, and local governmental agencies. This section discusses the market for U.S. coal exports, the regulatory framework governing approval of coal export facilities, and the difficulty coal companies have faced in building these facilities in part because of the ability of opponents to challenge such projects at the state and local levels. It also discusses the dominance of railways in transporting coal across the United States from extraction sites to power plants and the relationship between railways and coal export facilities.

A. U.S. Coal Export Trends

EPA air quality regulations, state clean air policies and renewable portfolio standards (RPSs) discussed in Chapter 4, and the availability of low-cost natural gas made available by hydraulic fracturing, have resulted in a significant decrease in demand for the use of coal to generate electricity in the United States beginning in the late 2000s.[108] As a news source explains, "[f]rom 2002 to 2012, coal exports from the United States more than tripled, to 127 million tons, according to the International Energy Agency,"[109] and 25 percent of U.S. coal exports in 2012 went to China.[110] China is the

---

[106] *See* Keith Goldberg, *EPA Urges FERC to Weigh Climate Impacts in LNG Reviews*, LAW360 (Jan. 20, 2016), http://www.law360.com/articles/748775/epa-urges-ferc-to-weigh-climate-impacts-in-lng-reviews.

[107] *See* EarthReports, Inc. v. Fed. Energy Regulatory Comm'n, 828 F.3d 949 (D.C. Cir. 2016) (holding that FERC was not required to consider indirect effects, including climate impacts, of increased natural gas exports as part of its environmental review associated with approving the Cove Point LNG facility in Maryland from an import terminal to an export terminal); Sierra Club v. Fed. Energy Regulatory Comm'n, 827 F.3d 36 (D.C. Cir. 2016) (same analysis with regard to approval of LNG export facility in Texas).

[108] *See* COLUMBIA CTR. FOR CLIMATE CHANGE LAW, CARBON OFFSHORING: THE LEGAL AND REGULATORY FRAMEWORK FOR U.S. COAL EXPORTS 1 (July 2011), http://www.law.columbia.edu/null/download?&exclusive=filemgr.download&file_id=59591.

[109] Peter Moskowitz, *U.S. Coal's New Focus on Exporting Leaves a Cloud of Dust Over Louisiana*, ALJAZEERA AMERICA (Jan. 24, 2014), http://america.aljazeera.com/articles/2014/1/24/coal-s-new-exporteconomyleavesacloudofdustoverlouisiana.html.

[110] *Id.*; *25% of U.S. Coal Exports Go to Asia, but Remain a Small Share of Asia's Total Coal Imports*, U.S. ENERGY INFO. ADMIN. (June 21, 2013), http://www.eia.gov/todayinenergy/detail.cfm?id=11791; *see also* BROCK R. WILLIAMS & J. MICHAEL

single biggest importer of U.S. coal, with smaller amounts going to Asia, Canada, Mexico, and Europe. Indeed, the prospects for coal exports looked very promising at this point, as reported in 2013:

> Whereas coal's role in the US economy has receded markedly and is projected to continued shrinking, its use in Asia—notably China and India—is expected to maintain its robust demand growth for at least the next 25 years.[111]

Anticipating falling domestic demand, coal companies and railroads began looking to overseas markets such as China and India, where electricity use was skyrocketing and environmental regulations were (and are still) relatively lax.[112] Moreover, with oil and natural gas prices still "relatively strong" in Europe, there was an increased demand for inexpensive U.S. coal for electricity use.[113]

However, in the first months of 2014, "coal prices declined steadily . . . in response to a combination of increased supply and lower import demand from China,"[114] and that trend continued through 2015.[115] In 2015, U.S. coal production fell to its lowest volume since 1986 as a result of less expensive sources of domestic alternative energy, stricter environmental regulations, and a continuing weak export market.[116] If the world coal market remains weak, the business opportunities for U.S. coal exports will likely continue to shrink, and coal exports may no longer be a lucrative endeavor.[117] China, the ultimate target market for global coal,

---

DONNELLY, CONG. RESEARCH SERV., U.S. INTERNATIONAL TRADE: TRENDS AND FORECASTS 31 (Oct. 19, 2012), https://fas.org/sgp/crs/misc/RL33577.pdf (stating that U.S. coal exports to China increased by a factor of 10 between 2009 and 2010).

[111] JOEL DARMSTADTER, RES. FOR THE FUTURE, THE CONTROVERSY OVER U.S. COAL AND NATURAL GAS EXPORTS 2 (Mar. 2013), http://www.rff.org/files/sharepoint/WorkImages/Download/RFF-IB-13-01.pdf.

[112] *Most U.S. Coal Exports Went to European and Asian Markets in 2011*, U.S. ENERGY INFO. ADMIN. (June 19, 2012), http://www.eia.gov/todayinenergy/detail.cfm?id=6750; U.S. DEP'T OF ENERGY, *supra* note 100, at Appendix B: Natural Gas Infrastructure, at 23 and Table B–6.

[113] Ed Crooks & Sylvia Pfeifer, *U.S. Coal Exports to Europe Soar*, THE FIN. TIMES (Oct. 3, 2012); U.S. ENERGY INFO. ADMIN., QUARTERLY COAL REPORT, DOE/EIA–0121 (2012/03Q) Table 7 (Dec. 20, 2012) (showing U.S. coal exports to Europe increasing nearly 30% between 2011 and 2012).

[114] Vicky Validakis, *The 2015 Energy Outlook Series: Coal,* AUSTRALIAN MINING (Dec. 8, 2014), https://www.australianmining.com.au/features/the-2015-energy-outlook-series-coal/.

[115] *See* Saqib Rahim, *With $1B Loss, Peabody Searches for Ways to Weather Industry Storm*, ENERGYWIRE (July 29, 2015) (reporting on weak global coal market that has caused dozens of smaller U.S. coal companies to declare bankruptcy and has also had major adverse financial impacts on larger coal companies such as Peabody, Arch Coal, Cloud Peak Energy, and Alpha Natural Resources which together supplied nearly half of U.S. coal in 2013).

[116] *Coal Production Falls to Lowest Volume in 3 Decades*, N.Y. TIMES (Jan. 8, 2016), http://www.nytimes.com/2016/01/09/business/coal-production-falls-to-lowest-volume-in-3-decades.html.

[117] Rahim, *supra* note 115.

announced in 2014 that it would cap its coal consumption by 2020.[118] At the end of 2014, the Institute for Energy Economics and Financial Analysis (IEEFA) issued a report concluding that given current market conditions, proposed coal export terminals did not make economic sense, and, moreover, it would be unprofitable for coal companies to use them even if they were constructed.[119] Nevertheless, U.S. coal companies continue to seek export opportunities, anticipating improved market conditions in the future.[120] But in the interim, several major coal U.S. coal companies filed for bankruptcy in 2016, creating additional uncertainty in the industry.[121]

B. U.S. Coal Transport and Regulation of Coal Export Facilities

While coal can be transported using barges, trucks, and even pipelines, the dominant method for transporting coal within the United States is rail.[122] Today, the greatest flow of freight anywhere in the United States consists of rail transport of coal from production sites to coal-fired power plants in the Midwest.[123] If the coal industry continues to shift towards an export-oriented business model, huge volumes of coal will need to be transported to West Coast ports for shipment to Pacific Rim purchasers.[124] Most of the rail tracks

---

[118] *China Sets Cap on Energy Use*, SHANGHAI DAILY (Nov. 19, 2014), http://www.shanghaidaily.com/article/article_xinhua.aspx?id=253473.

[119] *See* TOM SANZILLO, INST. FOR ENERGY ECON. & FIN. ANALYSIS, NO NEED FOR NEW U.S. COAL PORTS: DATA SHOWS OVERSUPPLY IN CAPACITY (Nov. 19, 2014), http://www.ieefa.org/wp-content/uploads/2014/11/Sanzillo-port-capacity.pdf.

[120] Adam Leach, *U.S. Coal: Could Exporting to Asia Offer a Lifeline to a Dying Industry?*, POWER-TECHNOLOGY.COM (Feb. 10, 2015), http://www.power-technology.com/features/featureus-coal-could-exporting-it-to-asia-offer-a-lifeline-to-a-dying-industry-4496432/; *see also Short-Term Energy Outlook, Coal*, U.S. ENERGY INFO. ADMIN. (July 7, 2015) (stating that U.S. coal exports are expected to increase in 2016 and remain at that level for at least two years); AM. ASS'N OF R.RS., RAILROADS AND COAL 9 (July 2015), https://www.aar.org/BackgroundPapers/Railroads%20and%20Coal.pdf ("U.S. coal producers are hopeful that coal exports will grow in the future, with Asia—especially China and India—seen as key potential markets.").

[121] *See* Clifford Krauss, *Peabody Energy, A Coal Giant, Seeks Bankruptcy Protection*, N.Y. TIMES (Apr. 13, 2016), http://www.nytimes.com/2016/04/14/business/energy-environment/peabody-energy-coal-chapter-11-bankruptcy-protection.html.

[122] Barges and pipelines are often cheaper than rail, but require huge flows of water not available in the American West and are vigorously opposed by railroad companies.

[123] *Railroad Deliveries Continue to Provide the Majority of Coal Shipments to the Power Sector*, U.S. ENERGY INFO ADMIN.: TODAY IN ENERGY (June 11, 2014), http://www.eia.gov/todayinenergy/detail.cfm?id=16651; *see also* QER, Chapter 5, Improving Shared Transport Infrastructure 5–9 (April 2015) ("Coal represents 39.5 percent of total tonnage moved by rail and 19.9 percent of total revenues for the railroad industry, and it is viewed by many in the industry as its most important single commodity."); AM. ASS'N OF R.RS., *supra* note 120, at 6 ("Coal is the most important single commodity carried by U.S. freight railroads.").

[124] W. ORG. OF RES. COUNCILS (WORC), EXPORTING POWDER RIVER BASIN COAL: RISKS AND COSTS (2011), http://www.worc.org/media/Exporting-PRB-coal-risks_and

carrying coal westward are currently at or near capacity, and would need to be upgraded to support this new traffic.[125] This demand for coal would also need to be met through the construction of new and expanded coal export terminals. Coal export terminals consist of physical facilities and related equipment necessary to unload coal from rail cars, stockpile, store, and blend the coal, and load it by conveyor onto ships for export.[126] These terminals require acquisition of significant amounts of industrial land in coastal port areas.

Although federal law does not restrict the export of coal to overseas markets, the recent effort to export coal produced in areas such as the Powder River Basin region of Montana and Wyoming to Asian markets requires new coal export facilities, which do require various governmental permits, leases, and approvals.[127] At the federal level, the U.S. Army Corps of Engineers must review and permit the siting and construction of coal export facilities in navigable waters under the Rivers and Harbors Act, the Clean Water Act, and the National Environmental Policy Act.[128] This evaluation requires a public interest review, evaluation of practicable alternatives, and consideration of the environmental effects associated with the facility and its impact on navigable waters.[129]

_costs-9-30-11.pdf (citing public statements by coal industry executives to arrive at an estimate of 140 million metric tons of coal bound for export).

[125] *See* CAMBRIDGE SYSTEMATICS, INC., NATIONAL RAIL FREIGHT INFRASTRUCTURE CAPACITY AND INVESTMENT STUDY PREPARED FOR ASSOCIATION OF AMERICAN RAILROADS (2007), http://www.nwk.usace.army.mil/Portals/29/docs/regulatory/bnsf/AAR2007.pdf.

[126] For instance, the permit application filed with State of Washington for the Millennial Bulk Coal Export facility near Longview, Washington described the project as using 100 acres of a 416 acre site and containing two docks, two shiploaders, four stockpile pads, one tandem rotary dumper, eight rail lines, and associated facilities, conveyors, and equipment capable of processing 44 million metric tons of coal per year. *See* Millennium Bulk Terminals Longview, LLC (MBTL) Coal Export Terminal Joint Aquatic Resources Permit Application (JARPA) Form at 6–7 and attached diagrams of proposed facility (Feb. 22, 2012).

[127] DARMSTADTER, *supra* note 111, at 4.

[128] *See* JAMES BACCHUS & ROSA JEONG, GREENBERG TRAURIG LLP, LNG AND COAL: UNREASONABLE DELAYS IN APPROVING EXPORTS LIKELY VIOLATE INTERNATIONAL TREATY OBLIGATIONS 3–4 (Nov. 2013) (prepared for Natl. Assoc. of Mfrs.); THE SIERRA CLUB, PERMITTING COAL EXPORT TERMINALS IN THE PACIFIC NORTHWEST—AN OVERVIEW OF REGULATIONS AND THE POLITICAL CONTEXT 3–7, https://content.sierraclub.org/creative-archive/sites/content.sierraclub.org.creative-archive/files/pdfs/100_121_CoalExport_PNW_Whitepaper_05_low_0.pdf (discussing applicable statutes and regulations).

[129] THE SIERRA CLUB, *supra* note 128, at 3–6; *see also* 33 U.S.C. § 1344 (West 2016) (CWA provisions governing permits for dredged or fill material into waters of the United States); 33 U.S.C. § 403 (West 2016) (Rivers and Harbors Act provision governing wharves, piers, excavations, filling, and other obstructions of navigable waters); 42 U.S.C. § 4332 (NEPA environmental review requirements on federal agencies); 33 C.F.R. §§ 320.1–.4 (general regulatory policies for Army Corps public interest review under the CWA); 33 C.F.R. §§ 230.1, *et seq.* (West 2016) (Army Corps regulations for implementing NEPA); All. to Save the Mattaponi v. U.S. Army Corps, 606 F. Supp. 2d 121 (D.D.C. 2009) (applying CWA public interest standard to

One area of controversy regarding the environmental review process under the National Environmental Policy Act is whether the Army Corps of Engineers must consider the broader impacts of coal exports and coal combustion in its evaluation of the environmental impacts of the project or whether it must merely consider the direct impacts of the export facility.[130] As of 2015, despite repeated protests from environmental groups, the agency has continued to consider primarily the direct effects of coal export projects and has declined to adopt a broader scope of review for these facilities that would include train traffic, impacts on global climate, and the cumulative effects of multiple coal export projects.[131]

States have jurisdiction over coal export facilities under both federal law (with states exercising delegated federal authority) and state law. With regard to state jurisdiction under federal law, "all coal export terminals . . . need [Clean Water Act] permits for their wastewater discharges as well as to control storm-water runoff."[132] Under Section 401 of the Clean Water Act, "[a]ny applicant for a Federal license or permit to conduct any activity including, but not limited to, the construction or operation of facilities, which may result in any discharge into the navigable waters," must apply for certification from the relevant state authority to ensure that the project will comply with state water quality standards and other aquatic resource protection requirements.[133] Conditions placed on Section 401 water quality certifications may extend beyond matters directly related the potential discharge, and all conditions imposed

---

challenge to Army Corps permit issued to build a reservoir on a creek); THE SIERRA CLUB, *supra* note 128, at 6 (discussing applicable statutes and regulations).

[130] Elizabeth Sheargold & Smita Walavalkar, *NEPA and the Review of Coal Export Projects*, CLIMATE LAW BLOG (Aug. 22, 2013), http://blogs.law.columbia.edu/climatechange/2013/08/22/nepa-and-the-review-of-coal-export-projects/.

[131] U.S. Army Corps of Engineers, Mem. for Record, NWS–2008–260, Pacific International Terminals, Inc. (May 9, 2016), http://www.nws.usace.army.mil/Portals/27/docs/regulatory/NewsUpdates/160509MFRUADeMinimisDetermination.pdf; U.S. Army Corps of Engineers, Mem. for Record, NWS–2011–325, BNSF Railways at 1–3 (July 3, 2013), http://www.nws.usace.army.mil/Portals/27/docs/regulatory/News/SCOPEMFRGATEWAYBNSF.pdf; *see also* Manuel Quiñones, *Army Corps Declines Broad Environmental Reviews for Export Terminals*, GREENWIRE (June 18, 2013), http://www.eenews.net/stories/1059983051.

[132] THE SIERRA CLUB, *supra* note 128, at 6.

[133] 33 U.S.C. § 1341(a)(2), (d) (West 2016). A state has four options when presented with a request for Section 401 water quality certification. It may grant the application, grant the application with conditions, deny the application, or waive the application. *Id.*; EPA Office of Wetlands, Oceans, and Watersheds, Clean Water Act Section 401 Water Quality Certification: A Water Quality Protection Tool for States and Tribes 9–11 (2010) [hereinafter "CWA 401 Handbook"] ("States and tribes are authorized to waive § 401 certification, either explicitly, through notification to the applicant, or by the certification agency not taking action. If action is not taken on a certification request, 'within a reasonable time (which shall not exceed one year),' the state or authorized tribe has waived the requirement for certification.").

by states automatically become conditions of the federal permit or license for which certification is sought.[134]

Coal export projects within coastal zones must also seek state certification under the Coastal Zone Management Act.[135] Activity that would have reasonably foreseeable effects on any land use, water use, or natural resources within a state's coastal zone must receive a consistency determination by the designated state authority.[136] With regard to air emissions, coal export facilities are "potential sources of significant amounts of particulate matter" and are subject to permitting under the Clean Air Act (permitting that the EPA has delegated to states); each export terminal therefore requires a Clean Air Act permit.[137]

As for state law, coal export facilities require a variety of state and local leases, permits, and approvals. For instance, in Washington State, as summarized by the Sierra Club, "any project that involves placing infrastructure on the bed of a river or in marine waters—which are owned by the state in trust for its citizens—requires a lease from the state Department of Public Lands,"[138] which has broad authority to grant, condition, or deny such leases based on a public interest review. Coal export projects may also require leases and permits from local port authorities and cities.

Despite the detailed federal and state regulatory process for obtaining approval of coal export terminals, companies have recently proposed to construct several coal export terminals, focusing largely on the Pacific Northwest.[139] But there has been growing opposition to U.S. coal exports as a result of their potential impact on global climate change, and most proposed export terminal projects have been abandoned or have failed to receive requested permits. In 2012 a company that had proposed a coal export terminal in Grays Harbor,

---

[134] PUD No. 1 v. Washington Dep't of Ecology, 511 U.S. 700, 712 (1994); 33 U.S.C. § 1341(d) (West 2016); CWA 401 Handbook, *supra* note 133, at 10.

[135] 16 U.S.C. § 1456 (West 2016). Under the CZMA, all coastal states have the authority to ensure that any federal agency activity within that state's coastal zone is consistent with its federally approved coastal management plan. *Id.*

[136] 15 C.F.R. §§ 930.1, *et seq.* (West 2016); *see also* COLUMBIA CTR. FOR CLIMATE CHANGE LAW, *supra* note 108, at 23 (discussing CZMA application to all federal permitting activity, including CWA S4 permits and Rivers and Harbors Act permits).

[137] THE SIERRA CLUB, *supra* note 128, at 6. The Clean Air Act grants states the authority to issue air permits within their jurisdictions that are consistent with the provisions of federal law.

[138] *Id.* at 4.

[139] *Washington Coal Export Project Proposals*, WASH. DEP'T OF ECOLOGY, http://www.ecy.wa.gov/geographic/EISprocess.html (describing the proposed Gateway Pacific Terminal and Millennium Bulk Terminals-Longview in Washington and the Coyote Island Terminal in Oregon); *Coal Ports*, W. INTERSTATE ENERGY BD., http://westernenergyboard.org/topics/spotlight/coal-ports/ (noting "a total of 5 proposed coal exporting terminals in Washington and Oregon," but not noting the Grays Harbor proposal).

Oregon announced that it would not move forward with the project,[140] and the Port of Coos Bay, Oregon, abandoned negotiations for a specific rail export project in 2013.[141] Also in 2013 Kinder Morgan abandoned plans to construct an export terminal at the Port of St. Helens, Oregon.[142] With respect to agencies' hesitance to approve the remaining proposed terminals, in September 2014 the Portland, Oregon District of the Army Corps of Engineers temporarily stopped reviewing the proposed Coyote Island export terminal at the Port of Morrow while awaiting a court decision on an Oregon state agency's denial of a state permit for the terminal.[143] And in 2016 the Army Corps halted permitting of the proposed Gateway Pacific Terminal in Washington State, citing to its impacts on tribal fishing rights.[144] The Millennium Bulk Terminal in Washington State remains under review, however.[145]

Although most proposed coal export terminals in the Pacific Northwest have been abandoned by project sponsors or denied by government agencies, coal export projects are under review in other parts of the country. As of January 2014, there were "seven proposals for new coal terminals in Louisiana, Alabama and Texas and seven proposals for terminal expansions in those three states" according to news sources.[146] These projects have also faced difficulty in part because of declining coal export markets and in part because of challenges in the permitting process.[147] In March 2014, the environmental groups Gulf Restoration Network, Louisiana Action Network, and Sierra Club filed a complaint in federal district court

---

[140] Ashley Ahearn, *Plans for Coal Export Terminal in Grays Harbor Abandoned*, NW. PUB. RADIO (Aug. 17, 2012), http://nwpr.org/post/plans-coal-export-terminal-grays-harbor-abandoned.

[141] Scott Learn, *Port of Coos Bay Coal-Export Proposal Ends After 18 Months of Work*, OREGONIAN (Apr. 1, 2013), http://www.oregonlive.com/environment/index.ssf/2013/04/port_of_coos_bay_coal-export_p.html.

[142] Scott Learn, *Kinder Morgan Drops Plans to Build Coal Export Terminal at Port of St. Helens Industrial Park*, OREGONIAN (May 8, 2013), http://www.oregonlive.com/environment/index.ssf/2013/05/kinder_morgan_drops_plans_to_b.html.

[143] *Coyote Island Terminal Regulatory Permit Application Review*, U.S. ARMY CORPS OF ENG'RS, PORTLAND DIST. (July 31, 2013), http://www.nwp.usace.army.mil/Media/News-Stories/Article/492571/coyote-island-terminal-regulatory-permit-application-review/.

[144] *Army Corps Halts Gateway Pacific Terminal Permitting Process*, U.S. ARMY CORPS OF ENG'RS, SEATTLE DIST. (May 9, 2016), http://www.nws.usace.army.mil/Media/NewsReleases/tabid/2408/Article/754951/army-corps-halts-gateway-pacific-terminal-permitting-process.aspx.

[145] *Permitting*, MILLENNIUM BULK TERMINALS, http://millenniumbulk.com/projects/permitting/.

[146] Moskowitz, *supra* note 109 (citing Al Armendariz of the Sierra Club).

[147] Manuel Quiñones, *Sierra Club Warns Against Texas Export Plans*, GREENWIRE (Mar. 9, 2012); Anthony Rome, *Fewer Lumps of Coal for Port of Corpus Christi this Year (And Probably Next Year Too)*, EAGLE FORD TEX. (Dec. 10, 2013), http://eaglefordtexas.com/news/id/313/fewer-lumps-coal-port-corpus-christi-year-probably-next-year/.

against United Bulk Terminals, which operates a coal transfer facility on the Mississippi River.[148] The groups alleged that the facility "violates the [Clean Water] Act by discharging coal and petroleum coke ('pet coke') into the Mississippi River without a permit."[149] Environmental groups also persuaded a Louisiana parish court to overturn a Louisiana Department of Natural Resources approval of a coastal use permit for an export terminal proposed by Ram Terminals.[150]

---

[148] Complaint, Gulf Restoration Network v. United Bulk Terminals (E.D. La., Mar. 18, 2014) (Case 2:14–cv–00608), http://www.tulane.edu/~telc/assets/complaints/03-18-14%20Complnt%20GRN%20v%20United%20Bulk.pdf.

[149] *Id.* at 1. *See also* Benjamin Alexander-Bloch, *Environmental Groups File Lawsuit Against Coal Export Facility in Plaquemines Parish*, TIMES-PICAYUNE (Mar. 18, 2014), http://www.nola.com/environment/index.ssf/2014/03/environmental_groups_file_laws.html.

[150] *Plaquemines Parish District Court Overturns Permit for Proposed Coal Terminal*, WDSU NEWS (Dec. 30, 2014), http://www.wdsu.com/news/local-news/new-orleans/plaquemines-parish-district-court-overturns-permit-for-proposed-coal-terminal/30461178; Katherine Bagley, *Losing Streak Continues for U.S. Coal Export Terminals*, INSIDE CLIMATE NEWS (Jan. 12, 2015), http://insideclimatenews.org/news/20150112/losing-streak-continues-us-coal-export-terminals.

Chapter 2

OIL AND GAS EXTRACTION

I. Introduction

Oil and natural gas are key components of the U.S. energy system. Approximately 35% of U.S. electricity is generated using natural gas,[1] and most vehicles are powered by oil that has been refined into gasoline or diesel fuel. Further, natural gas is a key feedstock for many manufacturing processes.[2] And a small yet growing number of vehicles—approximately 150,000 U.S. vehicles—are also powered by natural gas.[3]

The United States now produces a large amount of the oil and gas used to power domestic electricity plants, vehicles, and many other components of the energy system. This is largely due to special oil and gas extraction techniques developed in Texas in the 1990s. These techniques, which include drilling horizontally through rock formations that contain oil and gas and hydraulically fracturing these formations, are used to access specific types of formations called "shales" and "tight sandstones," which contain large quantities of oil and gas. Through horizontal drilling, oil and gas companies drill deep underground to a rock formation and then tilt the drill bit to drill laterally through the formation, sometimes for more than a mile. After drilling the well, these companies then hydraulically fracture it using a special fracturing technique called "slick water" fracturing. Slick water fracturing involves pumping large quantities of water and smaller quantities of chemicals down a well at high pressure. The water and chemicals exit the well and enter the rock formation around the well. Because the water-chemical mixture is injected at very high pressure, it cracks open the rock formation, exposing more surface area of the formation and helping to release oil or gas trapped tightly within the pores of the rock. "Proppants" like sand, which are injected along with the water and chemicals, prop open the fractures and allow oil, gas, or both to flow through fractures and into the well.

This Chapter first discusses the law that applies to all types of oil and gas development, including conventional development—

---

[1] *Nationwide, Electricity Generation From Coal Falls While Natural Gas Rises*, U.S. ENERGY INFO. ADMIN. (Oct. 7, 2015), https://www.eia.gov/todayinenergy/detail.cfm?id=23252.

[2] *Bulk Chemical Feedstock Use a Key Part of Increasing Industrial Energy Demand*, U.S. ENERGY INFO. ADMIN. (May 29, 2015), http://www.eia.gov/todayinenergy/detail.cfm?id=21432.

[3] *Natural Gas Vehicles*, U.S. DEP'T OF ENERGY, ALTERNATIVE FUELS DATA CTR, http://www.afdc.energy.gov/vehicles/natural_gas.html.

which does not involve horizontal drilling and hydraulic fracturing techniques—and then turns to unconventional development. It explores how federal, state, local, and regional government actors all play important roles in governing U.S. oil and gas extraction and the variety of issues addressed by oil and gas law, including efficient development of oil and gas, avoidance of too many property rights disputes, and environmental impacts.

II. Division of State and Federal Authority over Oil and Gas Development

The United States is somewhat unique among countries that have oil and gas resources because private individuals own the majority of oil and gas resources in the United States. In other countries, governments own most oil and gas and either rely on a state-owned company to produce oil and gas resources or enter into an agreement with a multinational oil company to develop these resources. This section introduces the federal and state laws that apply to government-owned and privately owned minerals in the United States.

The U.S. government owns and manages some minerals, including the minerals in oceans and other waters that are beyond the state-owned portion of these waters, and approximately 700 million acres of onshore minerals.[4] As we discuss in further detail later in this Chapter, the most important federal laws that apply to exploration and mineral extraction in offshore waters are the Submerged Lands Act (SLA)[5] and Outer Continental Shelf Lands Act (OCSLA),[6] which identify the offshore minerals that are owned by the federal government as opposed to the states, and how the federal government must manage and lease offshore mineral resources and distribute royalties. States, too, play a role in federal offshore mineral development. Under the OCSLA, the federal government must receive state recommendations regarding offshore leasing and the approval of offshore production, and it must implement these recommendations if they are reasonable.[7] Further, the federal government pays the neighboring states a certain percentage of revenues from leasing federal offshore minerals.[8]

For federally owned onshore minerals, the most important federal regulations include the Mineral Leasing Act (MLA)[9] and the

---

[4] *Oil & Gas Statistics*, U.S. DEP'T OF THE INTERIOR, BUREAU OF LAND MGMT, http://www.blm.gov/wo/st/en/prog/energy/oil_and_gas/statistics.html.

[5] 43 U.S.C. §§ 1301, *et seq.* (West 2016).

[6] 43 U.S.C. §§ 1331, *et seq.* (West 2016).

[7] 43 U.S.C § 1345 (West 2016).

[8] 43 U.S.C. § 1337(p)(2)(B) (West 2016).

[9] 30 U.S.C. §§ 181, *et seq.* (West 2016).

Federal Oil and Gas Royalty Management Act (FOGRMA),[10] which dictate the process for leasing these minerals and how royalties are to be distributed. Additionally, the Federal Lands Policy Management Act (FLPMA) provides guidance to the Bureau of Land Management (BLM) on how it must manage federal resources, including oil and gas resources, for the benefit of the public. The BLM, using its FLPMA and MLA authority, has issued a variety of regulations designed to mitigate the impacts of oil and gas development on BLM lands. As with federal offshore minerals, states also play an important role in regulating the development of minerals on onshore federal lands. States apply their own oil and gas regulations to oil and gas development on BLM lands, meaning that entities drilling wells on these lands must comply with both BLM and state regulations. Further, states receive approximately 49 percent of royalty revenues generated from the leasing of federal onshore minerals within their borders.[11]

Of the remaining minerals in the United States, which are owned by states, municipalities, individuals, corporations, and other entities, the entity that owns the minerals chooses whether or not to develop the minerals and receive royalty payments for these minerals. In other words, these entities exercise independent control over the minerals similar to the control that they exercise over all other types of property. However, federal and state regulations for the conservation of oil and gas and for environmental protection influence the development of minerals on municipal, state, and privately-owned lands.

The most important federal laws that apply to oil and gas development—including development of state or privately-owned minerals—include the Endangered Species Act (ESA),[12] Clean Air Act (CAA),[13] Clean Water Act (CWA),[14] Oil Pollution Act (OPA),[15] Safe Drinking Water Act (SDWA),[16] and Comprehensive Environmental Response, Compensation, and Liability Act (CERCLA).[17] Under the ESA and other acts that apply to wildlife,

---

[10] 30 U.S.C. §§ 1701, *et seq.* (West 2016).

[11] Statement of Greg Gould, Director, Office of Natural Resources Revenue, U.S. Dep't of the Interior, Before the S. Comm. on Energy & Natural Res., *Leveraging America's Resources as a Revenue Generator and Job Creator* 4 (July 22, 2014), http://www.onrr.gov/About/PDFDocs/FINA-SENR-ONRR-Testimony-on-Federal-Mineral-Revenues-(FINAL-07212014).pdf; 30 U.S.C. § 191.

[12] 16 U.S.C. §§ 1531, *et seq.* (West 2016).

[13] 42 U.S.C. §§ 7401, *et seq.* (West 2016).

[14] 33 U.S.C. §§ 1251, *et seq.* (West 2016).

[15] 33 U.S.C. §§ 2701, *et seq.* (West 2016).

[16] 42 U.S.C. §§ 300f, *et seq.* (West 2016).

[17] 42 U.S.C. §§ 9601, *et seq.* (West 2016).

such as the Bald and Golden Eagle Protection Act[18] and Migratory Bird Treaty Act,[19] any individual that could impact an endangered species or certain protected birds and their habitat must work with the Fish and Wildlife Service to mitigate these impacts.[20] For example, oil and gas operators—the entities responsible for the development of well sites and for drilling wells—are supposed to avoid operating equipment too close to eagles' nests.[21] Further, operators that need to withdraw water out of a stream for use in oil and gas development, or operators that propose to build well sites near streams that could send damaging soil and pollutants into streams, have to modify their activities if there are endangered aquatic species in that stream.[22]

Federal regulations also apply to air pollutants emitted from oil and gas development. Under the CAA, oil and gas operators must limit the emissions of certain pollutants from the actual well and from tanks and processing equipment near and away from the well.[23] CAA regulations also limit pollutants emitted from certain diesel equipment like bulldozers used to prepare oil and gas well sites,[24] and from trucks that carry materials like water, drilling and fracturing equipment, and fracturing fluids to well sites and wastes away from well sites.[25]

Regulations under the CWA, which address water pollutants from oil and gas development, prevent oil and gas operators from discharging wastes from oil and gas extraction into surface waters without first treating those wastes and receiving a CWA permit.[26] In

---

[18] 16 U.S.C. §§ 668, *et seq.* (West 2016).

[19] 16 U.S.C. §§ 703, *et seq.* (West 2016).

[20] 16 U.S.C. § 1538(a). (West 2016).

[21] U.S. FISH & WILDLIFE SERV., NATIONAL BALD EAGLE MANAGEMENT GUIDELINES 9, 12 (2007), https://www.fws.gov/southdakotafieldoffice/NationalBaldEagleManagementGuidelines.pdf.

[22] *See, e.g.*, Endangered and Threatened Wildlife and Plants; Determination of Endangered Status for the Sheepnose and Spectaclecase Mussels Throughout Their Range, 77 Fed. Reg. 14,914 (Mar. 13, 2012); Endangered and Threatened Wildlife and Plants; Endangered Status for the Diamond Darter and Designation of Critical Habitat, 77 Fed. Reg. 43,906 (July 26, 2012).

[23] 40 C.F.R. §§ 60.5375(a)(1)–(4) (West 2016).

[24] For the regulations, see 40 C.F.R. §§ 60.4200, *et seq.* ((Subpart IIII), Standards of Performance for Stationary Compression Ignition Internal Combustion Engines) (West 2016); 40 C.F.R. §§ 60.4230, *et seq.* ((Subpart JJJJ), Standards of Performance for Stationary Spark Ignition Internal Combustion Engines (standards applicable to engine manufacturers).

[25] Greenhouse Gas Emissions Standards and Fuel Efficiency Standards for Medium- and Heavy-Duty Engines and Vehicles, 76 Fed. Reg. 57,106 (Sept. 15, 2011).

[26] 40 C.F.R. §§ 435.50, 435.52 (West 2016); Memorandum from Linda Y. Boornazian, Director, Envtl. Prot. Agency, Clarification of Technology-based Sediment Toxicity and Biodegradation Limitations and Standards for Controlling Synthetic-based Drilling Fluid Discharges (Oct. 10, 2003), https://www.epa.gov/sites/production/

some cases these regulations prohibit these discharges altogether.[27] Under U.S. Environmental Protection Agency (EPA) regulations issued under the authority of the CWA in June 2016, operators are not allowed to send wastewater from hydraulically fractured shale wells to wastewater treatment plants.[28] Instead, they have to inject the waste underground in a well approved under the SDWA, recycle or reuse the wastewater, or find another approved disposal method.

An SDWA permit, called a Class II underground injection control permit, allows oil and gas wastes to be injected underground for disposal. Conditions in the permit require the person constructing and operating the disposal well to implement practices that will prevent the contamination of underground sources of drinking water.[29] Further, to address spills that can harm both surface waters and underground waters if the spills seep downward through soil, under the CWA and OPA any operators storing a certain minimum quantity of oil at a site must prepare a Spill Prevention, Control, and Countermeasure Plan that addresses how the operators will avoid and clean up spills.[30] The CWA and OPA also require immediate reporting to a national hotline of any oil spills that cause certain types of impacts or involve a minimum quantity of oil, and these laws impose liability on operators for the cost of cleaning up spills.[31] Somewhat similar to the CWA and OPA, CERCLA imposes liability on oil and gas operators for the clean-up of non-oil and gas substances that contaminate sites, such as hydraulic fracturing fluids or toxic substances used in the drilling process.[32]

Although several federal laws apply to oil and gas development on private, state, and federal lands, the majority of regulation of oil and gas development on private and state lands—and even some regulation on federal lands—comes from the states. Under the U.S. Constitution, states have certain "reserved" authority, meaning powers that are deemed to have always been held by the U.S. states.[33] These reserved powers include the power to regulate to protect public health, safety, and welfare, and states use these powers to regulate oil and gas extraction, as we discuss in more detail below. Further, states have extensive regulatory authority over oil

---

files/2015-06/documents/synthetic-based-drilling-fluid-discharges-clarification-memo_2003.pdf.

[27] 40 C.F.R. § 435.32 (West 2016).

[28] Effluent Limitations Guidelines and Standards for the Oil and Gas Extraction Point Source Category, 81 Fed. Reg. 41,845 (June 28, 2016).

[29] 42 U.S.C. § 300h(b)(1) (West 2016).

[30] 33 U.S.C. § 1321(j) (West 2016).

[31] 33 U.S.C. § 1321; 33 U.S.C. § 2702 (West 2016).

[32] 42 U.S.C. § 9601(14) (West 2016); 42 U.S.C. § 9607 (West 2016).

[33] U.S. Const. amend. X.

and gas development because the EPA, which implements most federal environmental laws, delegates much of its federal authority over oil and gas development to the states. In the oil and gas context, the most important authority delegated to states exists under the SDWA. As introduced above, the SDWA addresses the injection of substances underground, including the injection of water or carbon dioxide in order to displace oil and gas from rock pores and push that oil and gas toward the well, and the injection of oil and gas wastes far underground in order to permanently dispose of those wastes.[34] Many states have received delegated authority from the EPA to regulate both of these types of injection.[35] And in some cases, the EPA has discovered that states have not done an adequate job of exercising this delegated authority. For example, officials in California's SDWA program previously allowed the unlawful injection of oil and gas wastes into underground water aquifers—some of which appeared to contain water that could have served as drinking water.[36]

As we discuss further in "Unconventional Onshore Mineral Extraction," regional entities—groups of states acting together—also regulate certain oil and gas activities. Further, local governments play a very important role in regulating oil and gas development. Where they have not been preempted by states, many local governments establish zoning districts in which oil and gas development is or is not allowed, require companies to file monetary bonds before drilling wells so that there will be money available if damages occur, limit the times during which drilling and hydraulic fracturing may occur, and establish numerous other requirements for drilling and fracturing within local ordinances.

The following sections discuss the ownership and leasing of minerals, how state laws affect oil and gas rights and disputes between different mineral owners, and the use of state and federal law to address the environmental impacts of unconventional oil and gas development.

III. Onshore Mineral Ownership and Leasing

In addition to receiving regulatory permission to drill for oil and gas, a company that wants to extract minerals must obtain property

---

[34] *Class II Oil and Gas Related Injection Wells*, ENVTL. PROT. AGENCY, http://www.epa.gov/uic/class-ii-oil-and-gas-related-injection-wells#well_types.

[35] *Primary Enforcement Authority for the Underground Injection Control Program*, ENVTL. PROT. AGENCY, http://www.epa.gov/uic/primary-enforcement-authority-underground-injection-control-program.

[36] Letter from Jane Diamond, Director, Water Division, U.S. Envtl. Prot. Agency Region IX, to Jonathan Bishop, Chief Deputy Director, Cal. State Water Res. Bd. and Steven Bohlen, Oil and Gas Supervisor, Div. of Oil, Gas, and Geothermal Res., Cal. Dep't of Conservation (Dec. 22, 2014).

rights in those minerals. Some companies purchase minerals in fee—thus fully owning the minerals—but the most common way of obtaining property rights in minerals is to lease them.

A. Identifying Types of Mineral Ownership

In the United States, individuals own parcels of property that include the surface of the land, the minerals beneath the land, and limited rights to the use of the air above the land. The mineral estate may be severed from the surface. Therefore, if Marisa owns a parcel of property ("Greenacre," for example), she may "slice" off the ownership of the minerals beneath Greenacre and transfer the minerals to someone else. Marisa may do this by preparing a deed that transfers "the minerals in, under, and produced from Greenacre" to another individual, or a deed that uses similar language. Alternatively, Marisa may transfer away all of Greenacre but reserve to herself "the minerals in, under, and produced from Greenacre." Both of these actions will sever the mineral estate: the former action will transfer the minerals underlying Greenacre to someone else, and the latter will cause Marisa to continue to own the minerals, while someone else will own the surface.

There are several questions that arise when a deed transfers or reserves minerals. One question is the type of minerals that have been transferred or reserved. For example, sometimes a deed will transfer "oil, gas, and other minerals" underlying a property, and the person who owns the oil and gas might argue that he or she also owns *all* other minerals under that property. But the meaning of "other minerals" is often vague. If a legal dispute arises as to the meaning of the term, courts apply traditional methodologies ("canons of interpretation") for determining this meaning. Under the doctrine *ejusdem generis*, which provides that items in the non-exclusive portion of the list (such as "other minerals") must be similar in nature to the specifically-listed items in the list (such as "oil and gas"), a court might say that the category of "other minerals" only includes minerals that are liquid or gaseous, such as oil and gas, and not hard minerals like gravel or coal. Alternatively, under the "community knowledge" test, a court might ask whether at the time of the transfer of the minerals, members of the general public would have understood "other minerals" to include substances like gravel or coal.

Still other courts might ask whether extracting the minerals requires surface destruction like strip mining. Some courts assume that any minerals that require surface destruction to be extracted were not intended to be included in the category of "other minerals" that were transferred or reserved in a deed. Texas courts historically used the surface destruction test but have since chosen to use the community knowledge test and to list certain minerals that are presumed to be included in "other minerals" or not. However, the surface destruction test still applies to minerals that were severed when the Texas courts still followed that test.[37]

Pennsylvania courts have similarly set a clear rule for the meaning of the phrase "minerals" or "other minerals," at least when it comes to determining whether that phrase includes natural gas. Regardless of when "other minerals" were transferred in a deed, will, or similar instrument, the Pennsylvania courts assume that the phrase does not include natural gas unless the parties demonstrated a clear intent to include natural gas. Thus, in Pennsylvania, when a mineral owner reserved "one-half [of] the minerals and Petroleum Oils" beneath the property, the owner failed to reserve the natural gas under the property.[38]

Another question that arises when a deed or similar instrument transfers mineral ownership is whether the instrument has transferred to a new owner physical, possessory ownership in the minerals underlying property or rather a nonpossessory "royalty" interest. A royalty interest is the right to a certain portion of the minerals produced from a mineral estate, or to a certain portion of the value of the minerals, without having to pay for any of the costs of extracting the minerals. The difference between a mineral interest and a royalty interest is a very important distinction. Owners of actual minerals own the minerals underlying property, and this property interest is comprised of a large bundle of rights. Unless an instrument conveying mineral ownership indicates that some of these rights have been stripped away, mineral ownership includes the right to develop the minerals, convey the minerals away or lease them to a company that will develop the minerals, retain royalties and other lease benefits, and use the surface as is reasonably necessary to access the minerals. In contrast to mineral ownership, someone who only owns a royalty interest in minerals typically only has the right to a "cut" of the minerals or their value.

There are several types of royalty interests. One is the "nonparticipating royalty interest" or NPRI. "Nonparticipating" refers to the fact that the owner of the royalty interest does not participate in decisions about leasing or developing the minerals and does not receive any of the benefits associated with leasing minerals, such as bonuses. An NPRI is created when the owner of the minerals underlying Greenacre transfers to another party "a 1/8 royalty in minerals produced from Greenacre" or uses similar language. This means that whenever minerals are produced from Greenacre, the NPRI holder will receive a fractional share of the minerals—in this

---

[37] Friedman v. Texaco, 691 S.W.2d 586, 589 (Tex. 1985).
[38] Butler v. Charles Powers Estate, 620 Pa. 1, 4, 25 (2013).

case, 1/8 of the minerals, or 1/8 of their value—and will not have to pay for any of the costs of producing those minerals. An NPRI can also be created that is dependent on the mineral lease. For example, an NPRI can be created when the owner of the minerals underlying Greenacre transfers to someone else "1/8 of any royalty contained in any lease of Greenacre minerals." This is a "fraction of" NPRI, meaning that the holder of the NPRI receives a fraction of whatever royalty is included in the lease whenever the owner of the Greenacre minerals leases the Greenacre minerals to an oil and gas operator.

Another type of royalty interest is the "landowner's royalty." This is the royalty that is created when a mineral owner leases minerals to an oil and gas company, turning over all ownership of the minerals but reserving a bonus, other lease benefits, and a royalty, such as 1/8 of any minerals produced by the oil and gas company. As with the NPRI, the holder of the landowner's royalty does not have to pay for any of the oil and gas company's costs of producing oil and gas unless the lease expressly indicates otherwise.

In order to determine whether an instrument conveying minerals has conveyed mineral ownership or a royalty, courts look to the intent of the parties and apply various canons of construction when the intent is difficult to ascertain from the face of the instrument. A deed entitled "mineral deed" is more likely to be construed by a court as conveying minerals, and a "royalty deed" is more likely to be construed as conveying a royalty, but the instrument title and other labels are not determinative. Another device that the courts deploy to interpret the type of ownership created by a deed is to note the way in which individuals own minerals within a state. Some states treat mineral ownership as meaning that a mineral owner possesses the minerals as they sit underground—before they are extracted.[39] In these "ownership in place" states, if someone conveys the minerals "produced from" a property, and does not include the language "in and under," the courts are more likely to construe this conveyance as a royalty. This is because an instrument conveying the minerals "in, under, and produced from" a property in an ownership in place state seems to be conveying full ownership of the minerals as they sit underground, whereas a conveyance of merely the minerals "produced from" the property seems to convey only a "cut" of the minerals that happen to be extracted from the property.

In contrast, other states treat mineral owners as only possessing the minerals once they are extracted from the ground.[40] In these

---

[39] *See, e.g.*, Stephens County v. Mid-Kansas Oil & Gas Co., 113 Tex. 160, 167 (1923).

[40] Dabney-Johnston Oil Corp. v. Walden, 4 Cal.2d 637, 649 (Cal. 1935).

"exclusive right to take states," individuals who own minerals have the right to drill for the minerals and engage in other activities to extract the minerals, and they also have all of the other components of the "bundle or rights" described above. But mineral owners do not in fact "own" the minerals as a possessory, tangible object until they have extracted the minerals from the ground. So in exclusive right to take states, a conveyance of the minerals "produced from" a property can suggest that mineral ownership is being conveyed. Courts therefore must deploy other modes of analysis to glean the intent of the parties as to whether a mineral or royalty interest was conveyed.

The question of whether someone who owns minerals owns them "in place" or only after they are extracted also can impact certain other legal issues, such as how to properly secure an interest in the minerals. It also can affect whether or not the owner of minerals may file a trespass claim when someone interferes with the owner's mineral rights as those minerals sit underground (for example, if someone illegally drills into those minerals and extracts them). In some exclusive right to take states, mineral owners must argue conversion rather than trespass when someone illegally drills into their minerals.

One issue that is not impacted by the type of mineral ownership someone has (ownership in place versus exclusive right to take) is a common law principle called the "rule of capture." Regardless of whether a mineral owner owns minerals as they are sitting underground or only owns the right to extract the minerals, the rule of capture, which applies in all U.S. oil and gas producing states, provides that if someone legally drills a well on her property and that well happens to drain or "pull" away oil and gas from neighboring properties, the person does not owe the neighboring mineral owners any damages for the oil and gas that she drains away. In other words, she is permitted to extract as much oil and gas as possible from the well that she has drilled on her land, even if much of that oil and gas ends up being pulled from other, neighboring lands. The Texas Supreme Court determined in 2008 that under the rule of capture, an oil and gas company may even extend hydraulically-created fractures into neighboring minerals and drain gas without having to pay damages to the neighboring owners of gas.[41] A federal district court in West Virginia disagreed, finding that the rule of capture should not apply to formations like shales, which tightly trap oil and gas within pores and are not easily drained.[42]

---

[41] Coastal Oil & Gas Corp. v. Garza Energy Trust, 268 S.W.3d 1, 17 (Tex. 2008).

[42] Stone v. Chesapeake Appalachia, LLC, 2013 WL 2097397 at *6–*7 (N.D.W.V., Apr. 10, 2013) (case later settled and opinion vacated).

B. Oil and Gas Conservation Regulation

The rule of capture encourages oil and gas development by allowing those who are most motivated to drill to drain as much oil and gas as they wish from neighboring properties. But it also encourages inefficient production in at least two ways. First, numerous oil and gas companies might drill more wells than necessary, competing to be the first to drain neighboring properties. Second, oil and gas companies competing to be the first to drain the resources sometimes extract oil and gas too quickly. This can cause natural pressure in the underground formation from which oil and gas is produced—pressure that "pushes" the oil and gas up the well and prevents companies from having to pump oil and gas out of the well—to dissipate too quickly. It also can leave valuable oil and gas stuck within the rock pores.

States address the wasteful development of oil and gas that can result from the rule of capture through several types of statutes and regulations. First, through spacing rules, states require that in order to drill a well, an oil and gas producer must own—through mineral leases or direct fee simple ownership—a minimum amount of mineral acres. This prevents numerous wells from being drilled too close together. Some spacing rules are very simple. They apply statewide and provide that each well must be drilled a minimum distance from the property line, for example. Other spacing rules apply specifically to the formation being drilled into and hydraulically fractured. Based on exploratory wells that have been drilled into the formation and other data, the state determines the number of acres from which one well can efficiently produce oil and gas—in other words, the radius of drainage. The state then requires that for each well drilled into the formation, the well driller must own that minimum number of acres.

For example, say that a state determines that one well drilled into a formation produces oil and gas from 640 acres around the well. The state might establish a spacing order for that formation that requires that for anyone to drill a well, the person must have at least 640 acres of minerals. Therefore, if someone owns just 1 acre of minerals in the formation, the person cannot drill a well. Rather, the 1 acre of minerals must be "pooled" (combined with) at least 639 other neighboring acreages. One well is then drilled on the 640-acre area, and the proceeds are divided among the different mineral owners within that area. Pooling of acreage together to create minimum spacing units can be done in two ways. If an oil and gas company has permission to pool from mineral owners through a "pooling clause" in the lease, the company may pool together the minerals. Alternatively, if some mineral owners object to pooling, in many states there is "compulsory" or "forced" pooling. Oil and gas companies that have leased minerals, or mineral owners who want to produce the minerals themselves, may apply to the state to create a spacing unit. As long as a minimum percentage of mineral owners agree to pooling of the minerals, the rest of the objecting mineral owners are required to join the spacing unit, and all mineral owners receive a share of the proceeds from drilling based on their contribution of acreage to the unit and other factors.

A more sophisticated means of ensuring very efficient development of a formation is for a state to look at an entire formation, or a very large portion of a formation, and determine the exact number of wells, and the location of the wells, that would pull the most oil and gas from that formation. This is called "unitization," and it essentially involves pooling on a larger scale. Unitization often occurs when an oil and gas company that has leased minerals in a formation, or several of these companies, apply to the state oil and gas agency to request unitization. As with pooling, all mineral owners within the unit receive a proportionate share of proceeds based on their individually-owned acreage and other factors, such as the likely percentage of oil and gas that sits beneath their property.

Yet another means of addressing the wasteful development of oil and gas that can occur as a result of the rule of capture is to limit the amount of oil and gas that can be extracted from a given well or from a certain number of wells drilled into a formation. These limits, called "allowables" or "prorationing," are established by states and ensure that oil and gas is not extracted too quickly, and thus that oil and gas do not remain stuck in rock pores underground. Similarly, most states limit oil and gas companies' ability to release natural gas into the air through a process called "venting." Many oil wells produce natural gas that comes out of the well along with oil. This "associated gas" is sometimes viewed as a waste product by drilling companies, particularly if there are not pipelines nearby that can transport the gas to market. But states view venting as wasteful and environmentally detrimental. Instead of venting, some oil and gas companies flare (burn off) the natural gas that comes out of wells along with oil, which is also wasteful but sometimes necessary when pipelines are unavailable and the gas cannot be used on site. States sometimes limit operators to a certain amount of flaring,[43] and some states, like North Dakota, prevent the flaring of gas from oil wells that have operated for a least a year.[44] However, North Dakota

---

[43] OKLA. ADMIN. CODE § 165:10–3–15 (West 2016).
[44] N.D. CENT. CODE § 38–08–06.4 (West 2016).

regulatory officials often waive this requirement.[45] Further, in 2016, the EPA finalized federal regulations that limit the flaring of gas from newly-fractured oil and gas wells.[46]

C. Leasing of Privately-Owned Minerals

Once it is established that someone owns minerals, as opposed to a royalty interest, the owner of the minerals has two options if she wishes to extract the minerals. She may purchase or rent drilling and hydraulic fracturing equipment and produce the minerals independently, although this is a very expensive endeavor. Alternatively, and more commonly, mineral owners choose to have an oil and gas company produce the minerals. These owners lease the minerals to the oil and gas companies and reserve for themselves a portion of the value of the minerals through a landowner's royalty.

The oil and gas lease is not like a typical lease for a building. Rather, it conveys to the oil and gas company—the lessee—either a fee simple determinable interest in the minerals (the mineral owner retains the reversion interest) or a profit à prendre, which is the right to remove minerals from the land. A fee simple determinable is a property interest that potentially lasts forever, but certain events can cause the property interest to revert to the owner who originally conveyed away the fee simple determinable. Regardless of whether the lessee receives a fee simple determinable or a profit à prendre in the minerals, this is a very strong property right. The lessee obtains nearly all of the bundle of rights originally held by the mineral owner, including the right to explore for and produce the minerals, the right to use the surface as is reasonably necessary to extract the minerals, the right to transfer the lease to someone else (unless that lease contains restrictions on this type of transfer), and the right to retain profits from developing the minerals. In return for leasing away these strong property rights, the mineral owner-lessor retains several benefits, which include: (1) the bonus—a payment for agreeing to the lease; (2) delay rentals—payments that the oil and gas company makes during the initial term of the lease in order to avoid losing the lease despite not having yet drilled a well; and (3) a royalty.

The lease is not a permanent grant of mineral rights to an oil and gas company because it typically contains a primary term, during which the oil and gas company is supposed to start exploring for and drilling for oil and gas or else lose the lease. Further, most

---

[45] Mike Lee, *As Gas Continues to Go Up in the Air, N.D. Moves to Tighten Loose Flaring Rules*, ENERGYWIRE, Sept. 10, 2013, http://www.eenews.net/stories/1059986931.

[46] Oil and Natural Gas Sector: Emission Standards for New, Reconstructed, and Modified Sources, 81 Fed. Reg. 35,824 (June 3, 2016).

leases contain a secondary term, which is only triggered if the oil and gas company manages to start producing oil and gas by the end of the primary term. Primary terms typically last for five or ten years, and there are two types of primary terms that a lessor and lessee can bargain for. In one type, the oil and gas company works on a year-to-year clock. By the end of the first year, the oil and gas company must have commenced drilling a well or else pay delay rentals; otherwise the lease terminates. The same requirement applies through each subsequent year of the primary term. By the final year of the primary term, the operator typically must have started actually producing oil and gas or else it loses the lease. (Some leases contain limited exceptions, allowing operators who were in the process of drilling at the end of the primary term, or who had drilled recently but had not found oil and gas, and who continuously drill and eventually achieve production, to avoid losing the lease.)

In a second type of primary term—the paid-up lease—the operator pays an up-front amount of money to preserve the lease for the entire primary term and to avoid year-to-year requirements for drilling or paying delay rentals. But by the end of the paid-up term, this operator, too, typically must have commenced the production of oil and gas from the leased property or otherwise lose the lease.

If the operator under either a paid-up lease or a lease with delay rental provisions commences production by the end of the primary term, this carries the lease into the secondary term. The secondary term typically lasts "for so long as oil and gas is produced" from the leased acreage, or is produced "in paying quantities." As soon as production ends, the lease automatically terminates, and the minerals go back to the mineral owner unless the lease contains clauses that give the lessee a limited amount of time to re-start production without losing the lease. In some states, courts also apply certain common law remedies to avoid lease termination, but many strictly interpret the lease in favor of the lessor and thus in favor of termination. Termination of a lease between a lessor and a lessee that has not drilled for or produced oil and gas allows the lessor to lease to another company that might be more motivated to produce oil and gas from the property and generate royalties for the lessor.

D. Surface and Mineral Owner Disputes

As introduced above, the mineral portion of a property may be severed from the surface and sold separately, and minerals in many states have been severed from the surface. Where minerals have not been severed from the surface, the mineral owner, who also owns the surface, may lease minerals to an oil and gas company for development and include conditions in the lease that limit the company's use of the surface and prevent certain surface damages.

But when minerals have been severed, and the surface owner is not a party to the mineral lease, the surface owner has little recourse. This is because the common law in oil and gas producing states provides that the mineral estate is the dominant estate, and the surface estate is the servient estate. This means that the owner of the minerals, including the lessee who owns minerals by virtue of having leased them, may use the surface as is reasonably necessary to access the minerals and need not compensate the surface owner for any damages to the surface that occur as a result of this reasonable use. Courts have deemed numerous uses "reasonable." For instance, when oil and gas companies used up most of the surface owner's water,[47] or operated a waste pit directly next to the surface owner's house,[48] courts found that these companies owed no damages to the surface owner.

One common law doctrine softens this somewhat harsh rule for surface owners. This doctrine is called the "accommodation" doctrine. Under this doctrine, surface owners may be able to prevent certain surface uses by the mineral owner or lessee if they can show that they were using the surface for a particular use before the mineral owner or lessee entered the surface to begin drilling; that there are no reasonable alternatives to the surface use; and that the mineral owner or lessee has reasonable alternatives available, such as using different drilling equipment or changing the location of drilling, that would not destroy the existing surface use. If the surface owner can meet her burden of showing all of these things, the entity producing the minerals must deploy the demonstrated reasonable alternatives that avoid destroying or impacting existing surface uses.[49]

Further, a few states have enacted statutes, typically called surface damage acts, that require entities developing oil and gas to pay the surface owner even for damages caused by reasonable use of the surface. For example, in Oklahoma, before the producer of oil and gas enters the surface, the producer must attempt in good faith to negotiate with the surface owner regarding the damages that the producer will pay. If the producer and surface owner cannot agree on damages a court can appoint an independent appraiser to determine the damages amount, and if this still fails to result in agreement, the parties may request a jury trial.[50]

E. Leasing of Onshore Publicly-Owned Minerals

For onshore oil and gas owned by the U.S. government—most of which is on BLM lands—the oil and gas leasing and development

---

[47] Sun Oil Co. v. Whitaker, 483 S.W.2d 808, 812 (Tex. 1972).
[48] Grimes v. Goodman Drilling Co., 216 S.W. 202 (Tex. Civ. App. 1919).
[49] Getty Oil Co. v. Jones, 470 S.W.2d 618 (Tex. 1971).
[50] OKLA. STAT. ANN. tit. 52, §§ 318.5–318.6 (West 2016).

process is somewhat different. Congress through FLPMA, has directed the BLM "to manage public lands and resources in a manner that allows for multi-use development of lands, including 'a combination of balanced and diverse resource uses,' "[51] by current and future generations. Thus, when the BLM leases minerals, it is also supposed to consider other uses of BLM lands, such as recreation or grazing, that might come into conflict with mineral development. Further, the Mineral Leasing Act requires that the BLM, acting on behalf of the U.S. Secretary of the Interior, govern the surface-disturbing oil and gas activities "in the interest of conservation of surface resources"[52] and ensure "restoration of any lands or surface waters adversely affected by lease operations" by the operator.[53]

To plan for the potential leasing of oil and gas resources, the BLM conducts land use planning, looking at candidate areas for developing oil and gas resources on BLM lands and writing "resource management plans," or RMPs, for BLM areas deemed "open" to oil and gas development.[54] The RMP identifies potential impacts of oil and gas development on the area—particularly on sensitive portions of the area to be leased—and "*stipulations* needed to provide extra protection for sensitive resources in the plan area."[55] Once RMPs have been completed, any entity may nominate a parcel within an RMP for leasing, and the BLM may or may not decide to hold a competitive auction for that parcel, in which entities bid for leases in the parcel.[56] Anyone who obtains a lease within the parcel must follow the stipulations within the RMP for that parcel.[57] Once entities have obtained a BLM oil and gas lease they submit an application for a permit to drill (APD) to the BLM.[58] If the BLM believes that the drilling will have significant environmental impacts it must complete an Environmental Impact Statement (EIS) under the National Environmental Policy Act (NEPA) before issuing the permit, and if it believes that the impacts are not likely to be significant it must complete a shorter Environmental Assessment (EA) under NEPA. If the BLM decides to issue an APD, it often includes in its approval best management practices (BMPs) or conditions of approval (COAs) that require the lessee to engage in

---

[51] 43 U.S.C. § 1702(c) (West 2016).

[52] 30 U.S.C. § 226(g) (West 2016).

[53] *Id.*

[54] *Land Use Planning*, BUREAU OF LAND MGMT., http://www.blm.gov/wo/st/en/prog/energy/oil_and_gas/leasing_of_onshore/og_planning.html.

[55] *Id.*

[56] *Competitive Leasing*, BUREAU OF LAND MGMT., http://www.blm.gov/wo/st/en/prog/energy/oil_and_gas/leasing_of_onshore/og_leasing.html.

[57] *Id.*

[58] *Environmental Review and Permitting*, BUREAU OF LAND MGMT., http://www.blm.gov/wo/st/en/prog/energy/oil_and_gas/leasing_of_onshore/og_permitting.html.

certain practices that avoid or reduce environmental impacts.[59] If an operator receives an APD the operator may commence drilling, and the BLM periodically inspects the well site to determine whether the operator is following the BMPs and/or COAs within the permit and otherwise complying with BLM rules for drilling.[60]

One dispute that has recently arisen with respect to federal mineral leases is whether the royalties that the federal government receives for leasing federal minerals, which are owned in trust by the public, adequately cover the true costs of the minerals and the environmental impacts caused by extracting them. This type of "fair value" argument led the federal government to place a moratorium on federal coal leases in 2016, announcing that it would:

> launch a comprehensive review to identify and evaluate potential reforms to the federal coal program in order to ensure that it is properly structured to provide a fair return to taxpayers and reflect its impacts on the environment, while continuing to help meet our energy need.[61]

Environmental groups have called for similar evaluations of federal oil and gas leases,[62] but, aside from certain federal moratoria on offshore leasing discussed below, the federal government has not implemented a broad-based moratorium on federal oil and gas leases.

IV. Offshore Oil and Gas Leasing and Regulation

When oil and gas companies propose to drill a well offshore, the state or the federal government controls this offshore development. The entity with control over leasing and development depends on the location of the proposed offshore drilling. As introduced above, key statutes that establish federal and state jurisdiction over offshore minerals are the SLA[63] and the OCSLA. The SLA grants to states the title to certain submerged lands, including minerals that are part of those lands off of their coasts. Under the SLA, in all states other than Florida and Texas, the states own the minerals underlying waters that extend up to three nautical miles from their coastline.[64] For Texas and the western coast of Florida, these states own the minerals that extend out nine nautical miles from their coastline. The SLA

---

[59] *Id.*

[60] *Operations and Production*, BUREAU OF LAND MGMT., http://www.blm.gov/wo/st/en/prog/energy/oil_and_gas/leasing_of_onshore/og_production.html.

[61] *Secretary Jewell Launches Comprehensive Review of Federal Coal Program*, U.S. DEP'T OF THE INTERIOR (Jan. 15, 2016), https://www.doi.gov/pressreleases/secretary-jewell-launches-comprehensive-review-federal-coal-program.

[62] Bobby Magill, *Coal Moratorium Turns Spotlight to Oil, Gas Leases*, CLIMATE CENTRAL, Jan. 26, 2016, http://www.climatecentral.org/news/coal-moratorium-turns-spotlight-to-oil-gas-19957.

[63] 43 U.S.C. §§ 1301, *et seq.* (West 2016).

[64] 43 U.S.C. § 1312 (West 2016).

also makes clear that beyond the three and nine-nautical mile marks, the federal government owns submerged lands and their resources.[65] The OCSLA, in turn, describes the process that the federal government must follow in leasing and developing federal offshore minerals and applies various substantive restrictions on this leasing and development.

Congress has the authority to divide up jurisdiction over offshore minerals between states and the federal government—as it did in the SLA—due to several international agreements and executive orders. The 1982 United Nations Convention on the Law of the Sea codifies "customary international law" and establishes the boundaries within which nations control the waters, submerged lands, and air beyond their coasts. Specifically, the Convention provides that nations have sovereignty over their territorial sea, which nations may define as extending up to 12 nautical miles from the coastline.[66] The Convention also allows nations to claim certain sovereignty in "contiguous waters" and an "exclusive economic zone" (EEZ), both of which extend beyond their territorial sea. A nation's contiguous zone may extend up to 24 nautical miles from its coastline,[67] and its EEZ may extend up to 200 nautical miles from the nation's coastline.[68] U.S. Presidents subsequently established the specific boundaries of the U.S. territorial sea, contiguous zone, and EEZ, among other boundaries, through Presidential Proclamations. Further, Congress described the division of state and federal authority, and the specific federal authority that the government wields within these waters, through the SLA and OCSLA.

When an oil and gas company proposes to drill in state waters, it must first obtain a lease of the minerals from the state. However, some states have banned most or all leasing and development in state waters. For example, in Florida the state prohibits the issuance of permits to drill wells off of its east and west coasts.[69] It also prohibits structures for drilling for petroleum from receiving a permit or being constructed in state waters off of both coasts.[70] In states where drilling is allowed in state waters, after an oil and gas company obtains a lease of minerals from the state, it then applies to the state for a permit to drill.

For oil and gas development beyond state waters (in most states, beyond three nautical miles from the coast), the federal

---

[65] 43 U.S.C. § 1302 (West 2016).
[66] United Nations Convention on the Law of the Sea, arts. 2, 3, Dec. 10, 1982.
[67] *Id.*, art. 33.
[68] *Id.*, arts. 56, 57.
[69] FLA. STAT. § 377.24 (West 2016).
[70] FLA. STAT. § 377.242 (West 2016).

government—specifically, the Department of the Interior, which is responsible for federal lands—controls leasing and development. As with the states, many federal waters are off limits to oil and gas development. The OCSLA allows the President, through proclamation, to declare unleased federal waters off limits to leasing.[71] Further, by limiting budget appropriations to the Department of the Interior, which controls leasing, Congress can implement certain moratoria on leasing. And finally, Congress can directly declare offshore waters off limits to leasing. From 1990 through 2008, numerous federal waters off the coasts of California, Oregon, Washington State, parts of the North-, Mid-, and South-Atlantic, and parts of the Eastern Gulf of Mexico were off limits to federal leasing due to a Presidential moratorium.[72] George W. Bush rescinded this moratorium in 2008, with the exception of leasing in areas off of the coast designated as marine sanctuaries.[73] However, leasing in federal waters in parts of the Central and most of the Eastern portion of the Gulf of Mexico has remained off limits due to a Congressional moratorium implemented in 2006, which is still active. Specifically, through the Gulf of Mexico Energy Security Act of 2006, Congress made these areas off limits to leasing through 2022 but made other parts of the Gulf of Mexico open for leasing.[74] Further, in 2010 the "President withdrew Bristol Bay, offshore Alaska, from leasing consideration through June 30, 2017."[75] President Obama also canceled planned leases of federal minerals off of the Atlantic Coast in 2016.[76]

For the federally managed offshore minerals for which leasing is allowed, an agency within the Department of the Interior manages leasing and the issuance of various permits required for development. This agency was previously called the Minerals Management Service (MMS), but after the blowout of BP's Macondo oil well in the Gulf of Mexico in 2010 (the BP oil spill), the MMS was reorganized into three agencies. The primary agency now responsible for approving offshore leasing and development in federal waters is called the Bureau of Ocean Energy Management (BOEM). BOEM leases and permits development in four phases: planning for and

---

[71] 43 U.S.C. § 1341 (West 2016).

[72] ADAM VANN, CONG. RESEARCH SERV., OFFSHORE OIL AND GAS DEVELOPMENT: LEGAL FRAMEWORK 4–5 (Sept. 26, 2014), https://www.fas.org/sgp/crs/misc/RL33 404.pdf.

[73] *Id.* at 5.

[74] *Areas Under Moratoria*, BUREAU OF OCEAN ENERGY MGMT., http://www.boem.gov/Areas-Under-Moratoria/.

[75] *Id.*

[76] Jennifer A. Dlouhy, *Obama Bars Atlantic Offshore Oil Drilling in Policy Reversal*, Mar. 15, 2016, BLOOMBERG, http://www.bloomberg.com/news/articles/2016-03-15/obama-said-to-bar-atlantic-coast-oil-drilling-in-policy-reversal.

identifying offshore areas that will be leased through a five-year planning program; establishing dates and locations of auctions for leases and holding these auctions; granting requests by lessees to conduct exploration activities to ascertain the presence of minerals within leased areas; and issuing permits that allow for drilling and development.[77] Through the five-year planning program, BOEM maps out offshore areas for which it plans to issue leases after receiving public comments on the plan and prepares an EIS under NEPA.[78] Through the leasing process BOEM then auctions off specific parcels of minerals within these planning areas. Once an oil and gas company has obtained a lease, this does not give the company permission to conduct any oil and gas exploration and development. The company first must submit an exploration plan to BOEM that describes its plans to conduct initial exploratory work for the development of oil and gas, and BOEM determines whether a site-specific EA for the planned exploration is necessary under NEPA.[79]

BOEM also cannot approve exploration before notifying the state(s) that adjoin the proposed exploration and allowing those states to determine whether the proposed activities will be inconsistent with the state Coastal Zone Management Plan (CZMP) under the Coastal Zone Management Act (CZMA).[80] A CZMP is a plan, somewhat similar to a statewide land use plan for the coast, in which the state identifies important coastal and state water resources and prohibits or discourages certain activities that could impact those resources. If states declare that activities in federal waters will be inconsistent with their CZMP, the federal government may ultimately veto this finding and go ahead with the activity, but it first must undergo mediation and court review.[81] Following a finding of consistency under the CZMA (or a federal veto of a state finding of inconsistency), in approving exploration, BOEM attaches various requirements to the permit in an effort to prevent and mitigate environmental impacts.

If an oil and gas company decides to move beyond the exploration stage to drill wells that will produce commercial quantities of oil and gas, the company submits a plan for developing wells in the area(s) that it has leased. The company submits a document called a Development and Production Plan if the area in which it proposes to drill has not been significantly developed previously, and a Development Operations Coordination Document

---

[77] VANN, *supra* note 72, at 5.

[78] VANN, *supra* note 72, at 6–7.

[79] *Status of Gulf of Mexico Plans*, BUREAU OF OCEAN ENERGY MGMT., http://www.boem.gov/Status-of-Gulf-of-Mexico-Plans/.

[80] VANN, *supra* note 72, at 12.

[81] 16 U.S.C. § 1456 (c)(1)(B) (West 2016).

(DOCD) if the area has been previously developed.[82] A DOCD is "a plan that describes development and production activities proposed by an operator for a lease or group of leases."[83] BOEM must prepare an EIS under NEPA before approving these proposed plans or documents. Further, if a company proposes to hydraulically fracture a well or use other unconventional technologies anywhere offshore, or to drill in deep waters, the company also must submit a Deepwater Operations Plan and Conceptual Plan so that BOEM can address specific concerns relating to well safety.[84] And before drilling any well—even a well that is designed for the purposes of exploring an area to determine whether oil and gas is there—a company must submit an application for a permit to drill to BOEM and obtain approval.[85]

V. Unconventional Onshore Mineral Extraction

An increasingly common type of oil and gas development is "unconventional" extraction, which involves the use of hydraulic fracturing and other technologies to access difficult-to-extract minerals—many of which are located in very deep formations and are trapped tightly within tiny pores of rock. A large portion of the remaining oil and natural gas underlying the United States is within shale and tight sandstone formations, which have particularly small pores that are not well-connected to each other, thus impeding the flow of oil and gas from the formation. Technologies likely horizontal drilling and hydraulic fracturing must be used to produce oil and gas from these formations. This section expands upon the introduction to U.S. oil and gas law above, focusing on federal, state, regional, and local laws that apply specifically to unconventional oil and gas development and related activities, such as withdrawal of water for hydraulic fracturing and disposal of fracturing wastes.

As unconventional oil and gas development rapidly expanded in the United States beginning in the mid-2000s, more wells were drilled in more locations—including within states like Pennsylvania and North Dakota that had not recently experienced booms in oil and gas development. Thus, unconventional oil and gas techniques like horizontal drilling and hydraulic fracturing did not only introduce new environmental and social impacts; they also enabled the drilling of thousands of wells that otherwise would not␣been drilled, thus generating impacts simply due to the sheer number of new wells

---

[82] VANN, *supra* note 72, at 12.

[83] *Exploration Plan / Development Operations Coordination Document Approval Process and Definitions*, BUREAU OF OCEAN ENERGY MGMT., http://www.boem.gov/uploadedFiles/EP-DOCD_Facts_and_Definitions_BOEM.pdf.

[84] VANN, *supra* note 72, at 12.

[85] *Id.*

drilled. For example, Oklahoma experienced record numbers of earthquakes in 2014 and 2015—more earthquakes than occurred in California.[86] Many scientists have published papers that connect these earthquakes to the disposal of wastewater from drilled and fractured wells.[87] So while hydraulic fracturing itself is not creating this impact, hydraulic fracturing, by enabling the drilling of many new wells in Oklahoma's Woodford Shale and thus the production of large volumes of new wastewater, has contributed to the problem of "induced seismicity," which is the phenomenon of human-triggered earthquakes.

Legal regimes address these and other impacts of onshore oil and gas development in several ways. First, governments issue statutes and regulations that address oil and gas development. These oil and gas laws are numerous, but they include restrictions on the location of oil and gas development, regulations that address the technologies and techniques that may or must be used during development to lessen certain risks, permissible techniques for storing and disposing of wastes, information that oil and gas operators must disclose about their techniques and technologies used at well sites, and financial assurances that operators must provide to regulatory agencies in the event that impacts occur and the operators fail to fully address these impacts, thus requiring a public agency to incur costs to clean up pollution or otherwise fix a problem created by oil and gas development.

Second, when regulation fails to cover all of the impacts of oil and gas development, or operators fail to comply with regulations, common law causes of action are available for parties who wish to enjoin certain oil and gas activities or obtain damages for injuries caused by these activities. Numerous parties have filed lawsuits based on the common law of tort and property, often alleging nuisance, negligence, trespass, and assault (injury) when they are harmed by oil and gas activities.

Third, a large body of private law, including leases between oil and gas companies and mineral owners, contracts between oil and gas companies and lenders, and agreements among oil and gas

---

[86] Maria Gallucci, *Oklahoma Earthquakes 2015: Tremors Rise as Oklahoma Officials Struggle to Stem Fracking Wastewater Flow*, INT'L BUS. TIMES, Oct. 13, 2015, http://www.ibtimes.com/oklahoma-earthquakes-2015-tremors-rise-oklahoma-officials-struggle-stem-fracking-2138124.

[87] *See, e.g.*, OHIO DEP'T OF NAT. RES., PRELIMINARY REPORT ON THE NORTHSTAR 1 CLASS II INJECTION WELL AND THE SEISMIC EVENTS IN THE YOUNGSTOWN, OHIO, AREA 17 (2012), http://media.cleveland.com/business_impact/other/UICReport.pdf; Cliff Frohlich et al., *The Dallas-Fort Worth Earthquake Sequence: October 2008 through May 2009*, 101 BULL. OF THE SEISMOLOGICAL SOC'Y OF AM. 327 (2011); Katie M. Keranen et al., *Potentially Induced Earthquakes in Oklahoma, USA: Links Between Wastewater Injection and the 2011 M_w 5.7 Earthquake Sequence*, 41 GEOLOGY 699 (2013).

companies, addresses certain impacts of oil and gas development. This section briefly discusses these three types of laws.

A. Regulation of Unconventional Onshore Oil and Gas Development

When horizontal drilling and hydraulic fracturing became more common in the United States, states and the federal government began changing some of their statutes and regulations to address the new and expanded impacts associated with this development. For hydraulic fracturing on federal BLM lands, the BLM issued final rules in 2015 that focus primarily on information disclosure. Under these rules, before conducting a hydraulic fracturing operation, operators on BLM lands must collect and disclose to the BLM certain information about the area in which hydraulic fracturing will occur, such as the geology in the proposed area, existing faults and old wells, and usable-quality water near the proposed well.[88] Operators also must disclose how they plan to design the hydraulic fracturing operation, acquire water for hydraulic fracturing, and dispose of waste.[89] Further, operators must demonstrate prior to fracturing that the well, well casing, and cement that secures the casing in the well, are sound and can withstand the pressure placed on the well by fracturing.[90] After fracturing, operators must disclose the chemicals and volume of chemicals and water used in fracturing, the source from which they actually acquired water, and how they disposed of wastes from the well.[91] Operators may, however, claim that certain chemicals are trade secrets and avoid disclosing these chemicals to the BLM.[92]

All of this information can help BLM officials determine whether a well is likely to or did cause problems. For example, knowledge of the existing geology and faults in an area where a well is proposed to be drilled allows the BLM to ascertain whether there are existing fissures in the rocks surrounding the proposed well that could potentially serve as natural conduits through which fracturing fluids, natural gas, or oil that escaped an improperly-cased well could flow and could potentially cause pollution. Further, information about disposal practices could allow the BLM to determine whether any wastes were improperly disposed of on federal lands and could negatively impact these lands.

---

[88] Oil and Gas; Hydraulic Fracturing on Federal and Indian Lands, 80 Fed. Reg. 16,128, 16,218 (Mar. 26, 2015) (to be codified at 43 C.F.R. pt. 3160).

[89] *Id.* at 16,218–16,219.

[90] *Id.* at 16,219–16,220.

[91] *Id.*

[92] *Id.* at 16,220–16,221.

The BLM hydraulic fracturing rule also contains some substantive standards. For example, it requires that for most wells, wastes from hydraulic fracturing must be stored in tanks rather than pits dug on the surface of the well site.[93] Further, if data collected on the cementing and casing of the well shows that well integrity has been compromised and that substances like fracturing fluids and oil and gas flowing through the well could therefore potentially leak from the well, operators must take steps to remedy this problem.[94] However, certain oil and gas companies, several states, and the Ute Indian Tribe challenged the rule, arguing that the BLM lacks the authority to regulate hydraulic fracturing on federal lands and that the agency's rule is arbitrary. A federal district court in Wyoming temporarily stayed (prevented) BLM's enforcement of the rule while the cases are pending[95] and then invalidated the rule on the merits, finding that the BLM lacked the authority to regulate fracturing on federal lands.[96] The BLM, Sierra Club, and other groups have appealed this ruling.

States regulate oil and gas development on BLM lands in addition the BLM's authority, and some of these states believe that they already do an adequate job of regulating the impacts of development and that the BLM regulations are unnecessary. However, some of the BLM's new fracturing rules are more stringent than state rules and thus might provide more environmental protection on federal lands.[97]

For hydraulic fracturing on private lands, the federal government, regional government agencies (groups of states that work together), states, and local governments all play a role in regulating the environmental and social impacts of oil and gas development. States, however, play the primary role. Recently, many states have pushed back against what they view as "intrusions" into their oil and gas regulatory authority from both the federal government and local governments.

The federal government has addressed the specific impacts of hydraulic fracturing in several ways. The Fish and Wildlife Service has listed as endangered several species that are impacted by the

---

[93] *Id.*

[94] *Id.* at 16,220.

[95] Wyoming v. U.S. Dep't of the Interior, Order on Motions for Preliminary Injunction, Case No. 2:15–CV–043–SWS at 54 (D. Wyo., Sept. 30, 2015).

[96] Wyoming v. U.S. Dep't of the Interior, No. 2:15-CV-043, slip op. (D. Wyo., June 21, 2016).

[97] Hannah J. Wiseman, Written Testimony for "The Future of Hydraulic Fracturing on Federally Managed Lands," U.S. House of Representatives, Committee on Natural Resources, Subcommittee on Energy and Mineral Resources, July 15, 2015, http://docs.house.gov/meetings/II/II06/20150715/103846/HHRG-114-II06-Wstate-WisemanH-20150715.pdf.

development of well sites on which drilled and hydraulically fractured wells are located, in addition to water withdrawals from streams and other water resources to acquire the large volumes of water necessary to hydraulic fracturing.[98] For its part, the EPA has issued new CAA regulations that reduce the amount of volatile organic compounds (one of which is methane—a primary component of natural gas) that are released when fracturing fluids flow back out of wells,[99] and the EPA has directly regulated methane released from newly fractured and refractured oil and natural gas wells.[100] Further, as introduced above, the EPA has prohibited oil and gas companies from sending flowback water to wastewater treatment plants,[101] and it has issued non-mandatory guidelines for safely fracturing wells using diesel fuel.[102] And finally, the EPA has issued an advance notice of proposed rulemaking indicating that it might collect certain information on the chemicals used at fractured well sites under the Toxic Substances Control Act.[103]

There is also a major federal regulatory exemption, rather than regulation, for hydraulic fracturing. In 2005 Congress enacted a law that established that hydraulic fracturing does not count as "injection" under the SDWA.[104] Therefore, oil and gas operators that inject water and chemicals down a well for the purpose of hydraulically fracturing a well need not first obtain a SDWA underground injection control permit, which would require assurances that injection would not endanger underground sources of drinking water.

Some states have been very active in changing their regulations to address hydraulic fracturing. The most common regulatory change has been to require the disclosure of the chemicals used in hydraulic fracturing—a requirement that most oil and gas states have adopted. Most of these states allow well operators to keep confidential chemical information that operators deem to be trade secrets. At least two states—Idaho and Wyoming—substantively regulate the

---

[98] Kalyani Robbins, *Awakening the Slumbering Giant: How Horizontal Drilling Technology Brought the Endangered Species Act to Bear on Hydraulic Fracturing*, 63 CASE W. RES. L. REV. 1143 (2013).

[99] 40 C.F.R. § 60.5375(a)(1)–(4) (West 2016).

[100] *Supra* note 46 and accompanying text.

[101] *See supra* note 28 and accompanying text.

[102] Envtl. Prot. Agency, Revised Guidance: Permitting Guidance for Oil and Gas Hydraulic Fracturing Activities Using Diesel Fuels: Underground Injection Control Program Guidance #84 (Feb. 2014), https://www.epa.gov/sites/production/files/2015-05/documents/revised_dfhf_guid_816r14001.pdf.

[103] Hydraulic Fracturing Chemicals and Mixtures, 79 Fed. Reg. 28664, 28666–9 (May 19, 2014).

[104] *See* Hannah Wiseman, *Untested Waters: The Rise of Hydraulic Fracturing in Oil and Gas Production and the Need to Revisit Regulation*, 20 FORDHAM ENVTL. L. REV. 115 (2009) (discussing the law).

chemicals that may be used in fracturing, limiting the use of benzene, toluene, ethylbenzene, and xylene.[105] Further, a small group of states has banned or temporarily banned hydraulic fracturing, including Vermont, which has no significant oil and gas resources;[106] New York, which has a ban on fracturing using large volumes of water, at least for the duration of Governor Cuomo's leadership;[107] and Maryland, with a legislatively-enacted moratorium on fracturing through October 2017.[108]

Other states have updated regulations that address aspects of well development that can be impacted by the hydraulic fracturing stage (such as well lining ("casing"), which risks being compromised by the higher pressures of fracturing), or natural or human resources that can be similarly impacted. Several states like Colorado,[109] Michigan,[110] Ohio,[111] Pennsylvania,[112] and West Virginia[113] also require or strongly incentivize well operators to test the water quality in existing water wells near well sites before hydraulic fracturing. This produces "baseline data" that helps regulatory agencies and courts ascertain whether hydraulic fracturing or other stages of the well development process contaminated water in wells. Further, states like Texas updated their well casing regulations to ensure that wells can withstand the pressure placed on the well by hydraulic fracturing.[114] Pennsylvania made some of the most extensive changes to its oil and gas regulations, such as expanding the distance by which hydraulically fractured wells must be set back from natural resources like streams and wetlands;[115] requiring careful lining of pits that contain fracturing wastes and assurances that liners will not tear or be corroded by wastes in pits;[116] mandating

---

[105] IDAHO ADMIN. CODE r. 20.07.02.056 (02) (West 2016); Wyoming Rules and Regulations Oil General ch. 3 s 45(g) (West 2016).

[106] *Vermont First State to Ban Fracking*, CNN, May 17, 2012, http://www.cnn.com/2012/05/17/us/vermont-fracking/.

[107] Thomas Kaplan, *Citing Health Risks, Cuomo Bans Fracking in New York State*, N.Y. TIMES, Dec. 17, 2014, http://www.nytimes.com/2014/12/18/nyregion/cuomo-to-ban-fracking-in-new-york-state-citing-health-risks.html?_r=0.

[108] Josh Hicks, *Md. Fracking Moratorium to Become Law Without Hogan's Signature*, WASHINGTON POST, May 29, 2015, https://www.washingtonpost.com/local/md-politics/md-fracking-moratorium-to-become-law-without-hogans-signature/2015/05/29/e1d10434-062c-11e5-a428-c984eb077d4e_story.html.

[109] Colorado Oil & Gas Conservation Commission, Rule 609 Statewide Groundwater Baseline Sampling and Monitoring (2013), https://cogcc.state.co.us/COGIS_Help/SampleData.pdf.

[110] MICH. ADMIN. CODE r. 324.1002(3)(a) (West 2016).

[111] OHIO REV. CODE ANN. § 1509.06(A)(8)(c) (West 2016).

[112] 58 PA. CONSOL. STAT. ANN. § 3218 (West 2016).

[113] W.V. CODE § 22–6A–18 (West 2016).

[114] TEX. ADMIN. CODE § 3.13 (a)(4)(A) (West 2016).

[115] 58 PA. STAT. AND CONS. STAT. ANN. § 3215(b)(3) (West 2016).

[116] 25 PA. CODE § 78.56 (West 2016).

treatment of flowback sent to certain wastewater treatment plants;[117] and requiring that tanks containing chemicals must be placed within secondary containment (lined pits or containers that can catch leaking wastes). This containment must have the capacity to hold the volume of the largest tank plus ten percent of the volume.[118]

Finally, a few states, under their SDWA authority delegated from the EPA, have updated their regulation of disposal wells (underground injection control wells) that accept liquid wastes from hydraulically fractured wells and sometimes cause earthquakes. Ohio has some of the strictest regulations, requiring, *inter alia*, continuous monitoring of pressures around disposal wells and an automatic shut-off valve that will shut down the well if pressures appear to be too high and could cause nearby faults to slip, therefore potentially causing an earthquake.[119] Other states have not changed many of their oil and gas regulations in response to the boom in hydraulic fracturing and horizontal drilling.

Certain local governments, too, have been very active in regulating hydraulic fracturing, although states are increasingly preempting local governments from regulating. Some of the most extensive local regulations are found in Texas, where cities like Fort Worth and Arlington have lengthy oil and gas ordinances that limit the times during which fracturing (which is noisy) can occur and require oil and gas operators to purchase environmental liability insurance that will cover up to $5 million per incident, among many other requirements.[120] A Dallas, Texas ordinance required tracers to be used in fracturing so that any potential underground contamination could be tracked.[121] However, Texas has since preempted cities like Dallas from enacting this type of local ordinance. In 2015, after voters in Denton, Texas, voted to ban hydraulic fracturing in Denton,[122] the Texas Legislature preempted local governments from regulating most aspects of oil and gas development unless the governments had previously regulated fracturing for at least 5 years without impeding reasonable oil and gas development. The Texas Governor promptly signed this

---

[117] 25 PA. CODE § 95.10(b) (West 2016).

[118] 58 PA. STAT. AND CONS. STAT. ANN. § 3218.2 (West 2016).

[119] OHIO ADMIN. CODE 1501:9–3–7 (West 2016).

[120] ARLINGTON, TEXAS., ORD. No. 07–074, § 6.01(C)(2),(4); Fort Worth, Texas, Ord. No. 18449–02–2009, § 15–41(C)(4)(a).

[121] DALLAS ORDINANCE 29228 § 51A–12.201 (2013), http://citysecretary.dallascityhall.com/resolutions/2013/12-11-13/13-2139.PDF.

[122] Original Petition, Texas Oil and Gas Ass'n v. City of Denton, Cause No. 14–08933–431 (Nov. 5, 2014, Dist. Ct. of Denton County, Tex.), https://s3.amazonaws.com/static.texastribune.org/media/documents/TXOGA_Petition_file_stamped.pdf.

legislation into law.[123] Legislators in Oklahoma were also concerned about the local fracturing ban near the Oklahoma border and similarly preempted most local regulation of fracturing, again with a quick signature by the Governor.[124] Further, courts with jurisdiction in Colorado,[125] Ohio,[126] Louisiana,[127] and New Mexico[128] have declared that state law preempts most local control over oil and gas development or at least preempts bans or moratoria. Courts in New York[129] and Pennsylvania[130] have reached the opposite conclusion and have allowed local governments to maintain control over development.

Despite the trend toward preemption of local control over hydraulic fracturing, local governments in many states retain certain land use powers and other powers that they use to address the social impacts of hydraulic fracturing. For example, hydraulic fracturing requires hundreds of trucks to travel to and from well sites carrying fracturing materials and wastes,[131] and this heavy truck traffic can cause traffic congestion and damage roads. Many local governments have therefore entered into road use agreements with oil and gas operators in which the operators commit to repair or pay for repairs to damaged roads and in some cases commit to using certain routes. Further, local governments sometimes secure donations from the oil and gas industry, such as jaws of life for fire departments and police departments that have to respond to more emergency calls when more wells are drilled and more workers move into town. Garfield County, Colorado even formed a special committee to hear citizens' and others' complaints and concerns about oil and gas development and hydraulic fracturing within the county limits and to attempt to encourage industry to change practices in order to address these concerns.[132] The county has also installed air quality monitors

---

[123] Texas H.B. 40, http://www.legis.state.tx.us/tlodocs/84R/billtext/pdf/HB00040F.pdf (enrolled version) (signed by Governor May 18, 2015).

[124] Oklahoma S.B. 809 (enrolled version), http://webserver1.lsb.state.ok.us/cf_pdf/2015-16%20ENR/SB/SB809%20ENR.PDF (signed by Governor May 29, 2015).

[125] City of Longmont v. Colo. Oil & Gas Ass'n, 369 P.3d 573 (Colo. 2016); City of Fort Collins v. Colo. Oil & Gas Ass'n, 369 P.3d 586 (Colo. 2016).

[126] State ex rel. Morrison v. Beck Energy Corp., 143 Ohio St.3d 271, 275 (2015).

[127] Energy Mgmt. Corp. v. City of Shreveport, 467 F.3d 471, 475 (5th Cir. 2006).

[128] SWEPI v. Mora Cty., 81 F. Supp. 3d 1075 (D.N.M. 2015).

[129] Wallach v. Town of Dryden, 23 N.Y.3d 728, 755 (2014).

[130] Robinson Twp. v. Commonwealth of Pennsylvania, 623 Pa. 564, 696 (2013).

[131] U.S. DEP'T OF THE INTERIOR, NATIONAL PARK SERVICE, POTENTIAL DEVELOPMENT OF THE NATURAL GAS RESOURCES IN THE MARCELLUS SHALE 9 (2008), https://www.nps.gov/frhi/learn/management/upload/GRD-M-Shale_12-11-2008_high_res.pdf.

[132] *Energy Advisory Bd.*, GARFIELD CTY., http://www.garfield-county.com/oil-gas/energy-advisory-board.aspx.

around oil and gas sites and has conducted regular air quality monitoring.[133]

With all of the focus on federal, state, and local regulation of hydraulic fracturing, regional agencies are sometimes forgotten. But these agencies play an important role in regulating certain aspects of hydraulic fracturing or pre-fracturing stages—particularly water withdrawals for fracturing. In many parts of the United States, states, acting with the permission of Congress, have joined together to regulate the use of water from shared rivers. For example, in the Mid-Atlantic region the Susquehanna River Basin Commission (SRBC) and Delaware River Basin Commission (DRBC) regulate the use of water from their respective rivers based on consensus by the state representatives that are members of the Commissions. Any oil and gas operator who wishes to withdraw water from the watershed of the Susquehanna River must first obtain a permit from the SRBC, and within the permit the SRBC requires the operator to demonstrate that the water withdrawal will not harm aquatic species and requires monitoring of stream levels to ensure that stream levels do not drop below a certain point.[134] The SRBC also reserves the right to prevent previously approved water withdrawals; during a drought it temporarily prohibited numerous water withdrawals for fracturing, citing concerns that stream levels would drop too low.[135]

The Delaware River Basin Commission drafted regulations that would have applied to hydraulically fractured oil and gas wells within the watershed of the Delaware River, requiring things like pre- and post-drill testing of water resources and stormwater controls to prevent the runoff of pollutants during drilling and fracturing.[136] However, some of the DRBC's member states expressed concerns that the regulations were not adequately stringent, and there is currently a moratorium on fracturing within the watershed.[137]

---

[133] *Air Quality Management*, GARFIELD CTY., http://www.garfield-county.com/air-quality/.

[134] *See Frequently Asked Questions*, SUSQUEHANNA RIVER BASIN COMM'N, http://www.srbc.net/programs/natural_gas_development_faq.htm (describing monitoring protocols); *64 Water Withdrawals for Natural Gas Drilling and Other Uses Suspended to Protect Streams*, SUSQUEHANNA RIVER BASIN COMM'N, July 16, 2012, http://www.srbc.net/newsroom/NewsReleasePrintFriendly.aspx?NewsReleaseID=90 (describing passby flow regulations).

[135] *Id.*

[136] Del. River Basin Comm'n, Natural Gas Development Regulations (Nov. 8, 2011) (draft), *available at* http://www.nj.gov/drbc/library/documents/naturalgas-REVISEDdraftregs110811.pdf.

[137] E-mail from Kate Konschnik, Dir., Harvard Law School Envtl. Policy Initiative, to Hannah Wiseman (July 25, 2014).

B. The Common Law and Hydraulic Fracturing

When individuals believe that regulation has inadequately controlled the impacts of hydraulic fracturing or that oil and gas companies have flouted regulations and caused pollution or other problems, these parties have increasingly gone to court seeking to enjoin oil and gas activity or obtain damages. In some cases parties have alleged that chemicals and other hydraulic fracturing pollution migrated onto the surface and subsurface of their property, thus causing a nuisance or a trespass. Parties often file a nuisance claim and a trespass claim in the alternative when something like a chemical has entered their property. In some states courts strictly interpret a trespass as a physical invasion of property by a tangible object, and certain pollutants might not be considered to be physical invasions by these courts. Nuisances, on the other hand, involve activities that interfere within individuals' use and enjoyment of property and can include noise, air pollution, vibrations, and other non-tangible invasions of property. In other cases parties have argued that air pollutants and chemicals from fractured wells caused redressable injuries. In still other cases, individuals have alleged injury when injection disposal wells containing wastes from hydraulically fractured wells caused earthquakes, thus damaging homes and causing housing materials to fall on at least one inhabitant of a home.[138]

Many cases alleging common law claims relating to hydraulic fracturing have settled or have been unsuccessful[139] because parties alleging damages and injury from fracturing sometimes have difficulty proving that the damages and injury were in fact caused by fracturing—particularly where the alleged pollution is underground. However, there are at least two successful cases. In Texas, plaintiffs persuaded a jury to grant nearly $3 million in damages for a private nuisance caused by the air pollutant emissions from activities at the well site such as venting and flaring gas and using chemicals at the site.[140] Similarly, two families in Pennsylvania won several million dollars in a private nuisance verdict relating to alleged surface and groundwater pollution from drilling and fracturing activity.[141]

---

[138] Ladra v. New Dominion, LLC, 353 P.3d 529, 530 (Okla. 2015).

[139] *See* Arnold and Porter LLP, Hydraulic Fracturing Case Chart, http://files.arnoldporter.com/Hydraulic%20Fracturing%20Case%20Chart.pdf.

[140] Parr v. Aruba Petroleum, Cause No. CC–11–01650–E (County Court, Dallas County, Texas, Mar. 8, 2011).

[141] Ely v. Cabot Oil & Gas Corp., Civil No. 3:09–CV–2284, Verdict (M.D. Pa. Feb. 12, 2016).

C. Private Law Approaches to Controlling the Environmental Impacts of Hydraulic Fracturing

For entities concerned that regulations will not adequately control the impacts of hydraulic fracturing and associated activities or who simply wish to secure additional protections, several private law remedies are available. First, banks and other lenders that finance oil and gas development want to avoid exposure to borrower debt that could result from pollution of well sites, lawsuits associated with environmental damage, and other problems. These lenders therefore often attach conditions to lending agreements that require certain environmental protections at the well sites of their borrowers, such as prohibiting drilling and fracturing within a certain distance of water. Second, landowners who are lucky enough to own both the minerals and surface of property can include numerous environmental protections within the mineral lease if the oil and gas company-lessee agrees to these protections. For example, the landowner/lessor can require the operator to test water supplies on the property and provide results of the tests prior to drilling and fracturing, prevent the operator from bringing equipment on certain parts of the property, and require extensive restoration of the site after drilling and fracturing, among other lease provisions.

In addition to contracts that include provisions for environmental protection during oil and gas development, some oil and gas companies have voluntarily entered into memoranda of understanding with local governments, in which they agree to avoid using certain types of chemicals in hydraulic fracturing and to implement other environmental protections.[142] Further, at least one initiative to create voluntary best management practices—practices that exceed environmental regulations—is in progress. Through the Center for Sustainable Shale Development, oil and gas companies that operate in the Marcellus Shale region and environmental non-profit groups[143] created fifteen performance standards that the Center believes represent best practices, such as recycling 90% of produced water and fracturing flowback[144] and limiting the flaring (burning off) of gas from wells. Oil and gas companies may choose to follow these standards and to be certified by the Center as meeting the standards.

---

[142] *See, e.g., Memorandum of Understanding*, TOWN OF ERIE, COLO., AND ENCANA. See Encana, Erie Specific/Frequently Asked Questions at 2, http://www.erieco.gov/documentcenter/view/5781, for a brief description of the MOU.

[143] *About the Center for Sustainable Shale Development*, CTR. FOR SUSTAINABLE SHALE DEV., https://www.sustainableshale.org/about/.

[144] *Performance Standards*, CTR. FOR SUSTAINABLE SHALE DEV., https://www.sustainableshale.org/wp-content/uploads/2013/03/Performance-Standards-v.-1.4.pdf.

Fully describing the public and private standards that apply to hydraulic fracturing and other stages of oil and gas development would require a full book, but the regulations and private standards described above should provide readers with a clearer idea of the landscape of laws that exist.

Chapter 3

ENERGY TRANSPORTATION: ACCESSIBILITY, RELIABILITY, AND CYBERSECURITY

I. Introduction

The viability of the U.S. energy economy rests on the vast network of energy transport infrastructure that moves oil, gas, coal, and renewable energy resources from production sites to processing facilities, and then to wholesale and retail customers through an interdependent network of rail lines, electric transmission lines, oil pipelines, natural gas pipelines, and import and export facilities. As the Department of Energy explains, this energy transport system includes over 2 million miles of interstate and intrastate pipelines, more than 640,000 miles of high-voltage electric transmission lines, over 400 natural gas storage facilities, over 300 ports handling crude petroleum and refined products, and nearly 150,000 miles of railways that transport petroleum, gas, and coal.[1] This Chapter explores the federal and state regulation of energy transportation transactions, with particular emphasis on the laws governing (1) the regulation of electricity transmission; (2) the transportation of oil and gas by pipeline; and (3) the transportation of coal and oil by rail. These regulatory regimes are designed to counterbalance the potential for monopoly power in these industries and ensure that rates and terms are made available on a fair and nondiscriminatory basis to a variety of renewable and non-renewable energy providers. The laws discussed in this Chapter also address the safety and reliability of the energy transportation network. The siting and permitting of energy transport facilities are discussed in Chapter 1.

II. Regulation of Electricity Transmission

As explained in more detail in other chapters in this book, FERC regulates wholesale sales of electricity in interstate commerce and the transmission of electric energy in interstate commerce under Sections 201, 205, and 206 of the Federal Power Act (FPA). At the same time, states regulate retail sales of electricity within their borders and have primary authority over the siting and construction of electric transmission lines off of federal lands. Thus, even though states have primary authority over the actual construction and location of electric transmission lines, once those lines are in service,

---

[1] U.S. DEP'T OF ENERGY, QUADRENNIAL ENERGY REVIEW, SUMMARY FOR POLICYMAKERS, S–2 (Apr. 2015).

FERC exercises significant authority over the terms, price, and access that public utilities and other transmission line owners must provide to electricity generators and others wishing to transmit electricity across those lines. Using its authority under the FPA, FERC sets or approves rates for electricity transmission and acts to address undue preference or discrimination in the transmission of electricity and ensure that rates and charges are just and reasonable.

A. Transmission Access

FERC closely regulates the operation of transmission lines because, as explained in further detail in Chapter 6, operators of transmission lines can, and sometimes do, act in an anticompetitive manner when left to their own devices. Transmission line construction and operation is a capital-intensive, costly project, and only a limited number of entities are incentivized to take on these projects. Further, states, which permit the siting of transmission lines, do not want more transmission lines than are needed to transport electricity. FERC therefore regulates interstate transmission line operators to ensure that the somewhat limited number of operators in the business of building and constructing the lines provide access for generators that need to transport electricity to markets. Much of FERC transmission policy is geared toward encouraging the construction of more lines and opening up access to these lines in order to spur competition in electric generation, including a more diverse array of generators and a larger number of generators, which can lead to lower prices.[2]

One of the most important FERC transmission orders—also discussed in more detail in Chapter 6—is Order 888. This order requires line operators to provide line access for any generator who wants to use the line on a non-discriminatory basis, meaning that the rates charged must be reasonable, and the line operator may not charge itself more favorable rates than it charges of other users of the line. There are limits to open access, however. If large numbers of generators are already using the line and there is no room in the line for more electricity, or adding more electricity would compromise the reliability of the line's operation, the line operator need not allow the generator to use the line. Further, FERC cannot require transmission line owners or operators to build new lines in most cases, although it requires operators to *plan* for construction of new lines that may be needed.[3] When FERC issued Order 888 it also

---

[2] *See, e.g.,* Docket Nos. RM05–17–000 and RM05–25–000 at ¶ 5, 11 (Feb. 16, 2007) (noting that a FERC rule designed to encourage better planning for new transmission lines and more access to the lines would "facilitate[ ] the use of clean energy such as wind power" and would help to ensure that "efficient generating plants could obtain access to regional transmission grids," thus producing consumer benefits).

[3] *See infra* note 24.

issued Order 889, which requires transmission line operators to provide easily accessible, up-to-date information about the rates that it charges through an Open-Access Same-Time Information System (OASIS).[4]

In order to ensure that transmission line operators offer up their lines on an open access basis and do not charge unreasonable or discriminatory rates for the use of the lines, these operators must obtain a "tariff" from FERC before they operate a line. The term tariff, as used in the energy context, does not refer to a tax on goods. Rather, it is a license that allows a utility or other entity to charge a particular rate for the service that it offers (such as the use of a transmission line) and requires the entity to offer this service on certain terms—such as requiring open access to transmission lines.

One way in which FERC has encouraged more transmission lines to be built and has spurred further "opening up" of these lines to spur more competition in generation is by encouraging the formation of independent system operators (ISOs) and regional transmission organizations (RTOs). ISOs and RTOs are typically nonprofit organizations[5] that take over operational control of a particular set of transmission lines in a region from the individual owners of the lines within the region, although the line owners continue to maintain ownership of the lines themselves.[6] ISOs and RTOs are FERC-regulated entities and must receive a tariff from FERC before operating transmission lines. They are very important organizations—there are currently six RTOs and ISOs that are governed by FERC,[7] and they serve approximately two-thirds of electricity customers in the United States.[8]

To understand how RTOs and ISOs work, say that Utilities A, B, C, and D all own and operate transmission lines that cross various state lines in Region X. If an ISO or RTO is formed for Region X, this organization—with the permission of Utilities A, B, C, and D—will take over operational control of all of these utilities' wires, meaning that the RTO or ISO is responsible for (1) obtaining a tariff from FERC that establishes the rate that it may charge generators for use of the lines; (2) addressing and accommodating requests from new generators to interconnect with and use the wires; (3) ensuring that

---

[4] 75 FERC ¶ 61,078 (Apr. 24, 1996).

[5] 89 FERC ¶ 61,285 at 24 (Dec. 20, 1999).

[6] Alexandra B. Klass & Elizabeth J. Wilson, *Interstate Transmission Challenges for Renewable Energy: A Federalism Mismatch*, 65 VAND. L. REV. 1801, 1822–23 (2012) (describing RTOs and ISOs).

[7] *Regional Transmission Organizations (RTO)/Independent System Operators (ISO)*, FED. ENERGY REGULATORY COMM'N, http://www.ferc.gov/industries/electric/indus-act/rto.asp; *see also* Klass & Wilson, *supra* note 6, at 1822–23.

[8] *Electric Power Markets: National Overview*, FED. ENERGY REGULATORY COMM'N, http://www.ferc.gov/market-oversight/mkt-electric/overview.asp.

a constant voltage is maintained in the wires by balancing the amount of electricity "pulled" from the wires with the amount of electricity generated and sent through the wires; and (4) ensuring that the reliability of the lines is not compromised, both through long-term planning to ensure that there are adequate lines and shorter and medium-term tasks such as obtaining commitments from generators to build new power plants, which provide an adequate flow of electricity to meet future demand. One regional entity's operation of all of the wires within a region makes it easier for a variety of generators to send their electricity to customers throughout the region. Rather than having to deal with Utilities A, B, C, and D to transport their electricity across the region, these generators only have to work with the RTO or ISO. Some utilities also prefer giving operational control of their lines to an RTO or ISO because these utilities no longer need to receive an individual tariff from FERC in order to operate their wires. Rather, the RTO or ISO obtains the tariff, charges generators for the use of the lines following the rates set out in the tariff, and reimburses the utility owners of the lines.

FERC encouraged the formation of ISOs in Order 888 in 1996, setting out the requirements that organizations had to meet in order to be approved by FERC as an ISO. Specifically, FERC required, inter alia, that the organization's governance "be structured in a fair and non-discriminatory manner," including that the organization be "independent of any individual market participant or any one class of participants (e.g., transmission owners and [electricity] end-users)."[9] Further, the order provided that ISOs "should (1) have no financial interest in the economic performance of any power market participant"; (2) provide open-access, non-discriminatory transmission service under one "grid-wide tariff"; (3) "have primary responsibility in ensuring short-term reliability of grid operations"; (4) control all interconnected wires within the region covered by the ISO; (4) accommodate electricity "transactions made in a free and competitive market"; and (5) "promote the efficient use of and investment in generation, transmission, and consumption, among other requirements.[10]

By 1999, FERC had decided that it needed to do more to encourage the formation of ISO-type entities, and, in turn, to provide competing generation companies with easier access to a regional grid. Specifically, FERC concluded that since issuing Orders 888 and 889, it had determined that the Commission "must take further action if we are to achieve the fully competitive power markets

---

[9] 75 FERC ¶ 61,080 at 280 (Apr. 24, 1996).
[10] *Id.* at 280–85.

envisioned by those orders."[11] FERC therefore issued Order 2000, which encouraged the formation of an entity called an RTO. RTOs are essentially the same as ISOs, and in Order 2000 FERC—as it had done for ISOs in Order 888—specified the "minimum characteristics and functions that a transmission entity must satisfy in order to be considered an RTO."[12] These characteristics are largely the same as those required of ISOs, with slight modifications.[13]

The six FERC-governed ISOs and RTOs (called "RTOs" for simplicity) that operate today cover the Northeast and Mid-Atlantic regions and most of the Midwest. Transmission lines in the Southeast and West generally are not governed by these organizations, with the exception of the one-state ISO that operates in California.[14] The members of RTOs typically include utilities that are transmission line owners, "independent transmission companies" that own transmission assets but do not have retail electricity customers like utilities, entities that generate power and arrange for the sale and purchase of power ("power marketers"), and utilities that purchase electricity, among other members.[15] These members do not run the organizations but have important stakeholder input within the RTO.

Although RTOs now cover a good portion of the electricity-consuming population, FERC remains concerned that companies that own transmission lines—including companies that own the lines and transfer operational control over an RTO—still lack adequate incentives to build new lines. After all, constructing new transmission lines would simply create more opportunities for a variety of competing generators to use the lines.[16] FERC has therefore issued several orders that require transmission line owners and operators to plan for grid expansion to support new generation. The hope is that although FERC does not govern the siting of most transmission lines, this planning will help to encourage new construction of lines and incentivize states and local governments to allow some of these lines to be sited. In 2007 FERC issued Order 890, which, among other goals, aimed to "increase transparency in the rules applicable to planning and use of the transmission system."[17]

---

[11] 89 FERC ¶ 61,285 at 9 (Dec. 20, 1999).

[12] *Id.* at 1.

[13] *Id.* 151–322.

[14] *Regional Transmission Organizations (RTO)/Independent System Operators (ISO), supra* note 7; Klass & Wilson, *supra* note 6, at 1822.

[15] *See, e.g., Membership List,* MIDCONTINENT INDEP. SYS. OPERATOR, https://www.misoenergy.org/StakeholderCenter/Members/Pages/MembershipList.aspx.

[16] *See, e.g.,* 89 FERC ¶ 61,285 at 35 (Dec. 20, 1999) (noting that utilities that own electric generation, transmission, and distribution facilities still have incentives to "use their transmission assets to favor their own generation").

[17] Docket Nos. RM05–17–000 and RM05–25–000 (Feb. 16, 2007).

As summarized by FERC, this order required all transmission providers (operators of transmission lines) to "participate in a coordinated, open and transparent planning process on both a local and regional level."[18] Through this process, providers had to "meet with all of their transmission customers and interconnected neighbors to develop a transmission plan"[19] through meetings that were transparent and "open to all affected parties."[20] Providers also had to share information, such as where additional electric service (and thus new transmission lines) might be needed at specific locations.[21] The plan that resulted from this process was supposed to identify potential new or expanded lines that would meet "specific service requests" of those who use the lines,[22] among other factors. Transmission providers were also expected to share their plans with providers in neighboring systems to identify where there might be too much congestion in shared wires and to consider potential "system enhancements" to address this problem.[23] FERC expanded these transmission planning requirements in Order 1000, issued in 2011, which required each transmission provider to participate in a regional transmission planning process and to produce a transmission plan; identify in its tariff how the provider would consider transmission needs associated with "public policy requirements," such as state renewable portfolio standards that encourage more renewable generation and thus require more transmission lines; and expand "coordination between neighboring transmission planning regions," among other mandates.[24]

Order 1000 also encouraged the construction of more transmission lines by "merchant" owners—companies that are not large utilities that own generation, transmission, and distribution facilities but are rather independent entities that are solely in the business of building transmission lines and do not receive any cost recovery for their investment from ratepayers. Order 1000 encouraged more of these lines by removing certain impediments to these lines. Specifically, prior to Order 1000, when transmission plans identified the need for new transmission lines, transmission providers within the plans had to give an incumbent utility (acting with the approval of the federal government) the first opportunity to construct these lines.[25] This practice, called the "federal right of first

---

[18] *Fact Sheet: Order No. 890*, FED. ENERGY REGULATORY COMM'N, https://www.ferc.gov/industries/electric/indus-act/oatt-reform/order-890-fact-sheet.pdf.

[19] Docket Nos. RM05–17–000 and RM05–25–000, *supra* note 17, at ¶ 445.

[20] *Id.* at ¶ 460, 471.

[21] *Id.* at ¶ 486.

[22] *Id.* at ¶ 494.

[23] *Id.* at ¶ 523.

[24] 136 FERC ¶ 61,051 at 1 (July 21, 2011).

[25] *Id.* at ¶ 225.

refusal,"[26] pushed out potential, competing lines, including merchant lines. Order 1000 eliminated this federal right of first refusal, although it so far has not prohibited similar rights of first refusal required by state law.[27] To further encourage the construction of merchant generation, in 2013 FERC issued a policy statement that made it easier for entities wanting to construct a merchant transmission line to bargain directly with potential users of this line.[28] Specifically, these merchant companies could enter into bilateral contracts with potential line users (for example, generators) in which the merchant company and the line users agreed on the rate that the user would pay for the transmission line and the terms and conditions of the contract. This gave merchant providers more assurance—prior to constructing a line—that they had a viable business project with committed customers.[29] FERC does not automatically approve these negotiated rates, however. It looks to the following four factors to determine whether the rates are acceptable:

> (1) the justness and reasonableness of rates [whether the rate agreed upon is too high]; (2) the potential for undue discrimination; (3) the potential for undue preference, including affiliate preference [the threat of the merchant company favoring its affiliate companies when it charges them for use of the line]; and (4) regional reliability and operational efficiency requirements.[30]

The rate that operators may charge for the use of transmission lines has not just been an issue with merchant generators. It is also a very contentious issue in the RTO context, where the RTO often plans for the construction of new lines and, with FERC approval in the form of a tariff, allocates the costs of the line among the generators and utilities who use the line. Often, owners of generators and utilities who feel that they do not benefit from the new transmission line argue that the rates that they pay for the use of transmission lines within the RTO should not incorporate the costs of these new lines, as discussed in the following section.

B. Transmission Cost Allocation

As utilities and RTOs engage in more regional transmission planning, the issue of allocating the cost of new, regional transmission lines among RTO members has become increasingly

---

[26] *Id.* at ¶ 229.

[27] *Id.*; Alexandra B. Klass & Jim Rossi, *Revitalizing Dormant Commerce Clause Review for Interstate Coordination*, 100 MINN. L. REV. 129, 193 (2015).

[28] 142 FERC ¶ 61,038 (Jan. 17, 2013).

[29] *Id.* at 1–2; *see also* Klass & Wilson, *supra* note 6.

[30] Chinook Power Transmission, LLC, 126 FERC ¶ 61,134 at P 37 (Feb. 19, 2009).

important. Under the FPA and applicable case law, the guiding principle is "cost causation," meaning that "approved rates [must] reflect to some degree the costs actually caused by the customer who must pay them."[31] In a 2009 decision, *Illinois Commerce Commission v. FERC (ICC I)*, the U.S. Court of Appeals for the Seventh Circuit held that FERC's approval of one RTO's cost allocation scheme for new, high voltage transmission lines in that region did not satisfy the cost causation principle because it did not sufficiently establish that the benefits utilities would receive from those lines were "roughly commensurate" with the costs imposed on those utilities.[32] Within this RTO—the PJM RTO—cost allocation issues were particularly thorny because the RTO spans a wide region that runs from the Midwest to the mid-Atlantic states, and utilities in the Midwest felt that they were not receiving benefits from new, high-voltage transmission lines built farther to the east, which were designed to alleviate congestion. FERC believed that this would benefit the whole region because there would be fewer power failures throughout the region, but the midwestern utilities disagreed. The Seventh Circuit sided with the midwestern utilities, finding that FERC had failed to provide adequate numbers demonstrating how, and by how much, the midwestern utilities would benefit from the eastern lines.[33]

A few years later, however, in 2013, in *Illinois Commerce Commission v. FERC (ICC II)*, the same court upheld FERC's approval of another RTO's cost allocation approach for the construction of new high-voltage "multi-value" transmission lines designed to bring wind energy from wind-rich regions in the states in the northwestern portion of the RTO (such as North Dakota and Iowa) to population centers south and east.[34] In that case, which involved the Midcontinent Independent System Operator (MISO),[35] the court stressed the need for states to meet state renewable portfolio standards, the importance of integrating more renewable energy into the transmission grid, and the fact that MISO did enough calculation of benefits to be gained by those utilities challenging the costs to meet the cost causation principle.[36] The court reasoned that all utilities would have access to cheaper wind power as a result of the new lines and that FERC had provided numbers showing the smaller amount of electricity that would be lost from lines due to the

---

[31] Illinois Commerce Comm'n v. Fed. Energy Regulatory Comm'n, 576 F.3d 470 (7th Cir. 2009) (internal quotations omitted).

[32] *Id.* at 477.

[33] *Id.*

[34] Illinois Commerce Comm'n v. Fed. Energy Regulatory Comm'n, 721 F.3d 764, 780 (7th Cir. 2013).

[35] MISO is officially a regional transmission organization despite being called an independent system operator. (It was originally an ISO.)

[36] *ICC II*, 721 F.3d at 774–75.

electricity having to travel shorter distances over new lines.[37] Further, as a result of the multi-value projects, utilities would have to maintain less back-up generation for periods of peak demand because they would have access to more generation throughout the region, and FERC calculated the savings that utilities throughout the region would accordingly achieve.[38]

C. Transmission Interconnection

Just as FERC has attempted to enhance competition in electricity generation by encouraging regionally-operated grids, transmission planning, and new cost allocation techniques for the construction of transmission lines, FERC has also attempted to provide generators with easier grid access by improving its policies for interconnection. All electricity generators must connect to a transmission line in order to send their electricity to customers, and those building new generation plants often have trouble accessing lines. Before any generator may connect to the grid it is required to prepare a series of expensive studies and simulations to show that it will not add more electricity to the grid than the particular wires can handle and that it will not otherwise cause problems with grid reliability. These studies are expensive and take years to complete. Existing lines are often already too congested—thus allowing line operators that must follow open access policies to deny access to the grid. Long "queues" of generators waiting to obtain access to the grid have therefore formed. For example, at the end of 2007, calculations by MISO suggested that if the same queue system stayed in place, and interconnection requests continued to be granted at the same rate, all of the projects in the queue would not be processed until approximately 2050.[39] The delays were similar in other RTOs. More recently, as of July 2013, there were 23,198 megawatts (MW) of proposed generation in the MISO interconnection queue representing 138 projects. Of those projects 64 percent of the MWs and 86 percent of the projects were wind-related.[40]

In addition to the queue problem, as large amounts of new renewable generation came online in recent years, transmission line operators placed relatively onerous requirements on these generators before they could interconnect.[41] Transmission line

---

[37] *Id.* at 774.

[38] *Id.*

[39] *See* K. PORTER, ET AL., GENERATION INTERCONNECTION POLICIES AND WIND POWER: A DISCUSSION OF ISSUES, PROBLEMS, AND POTENTIAL SOLUTIONS 16 (Nat'l Renewable Energy Lab., Jan. 2009).

[40] *See* Randy Rismiller, *Grid Interconnection of Renewable Generation*, Nat'l Ass'n of Regulatory Utility Comm'rs (Aug. 14, 2013) (PowerPoint presentation).

[41] *See* 111 FERC ¶ 61,353 at ¶ 1 (June 2, 2005) (noting "unnecessary obstacles" to wind generators' ability to interconnect to the grid).

operators were concerned that due to the intermittency of renewable resources, which have electricity production surges on windy or sunny days and sudden drops in generation at certain times, these resources would interfere with the reliability of the grid. FERC has issued orders and guidance to address this concern and to address long interconnection queues.

With respect to both queue management and improved, predictable interconnection requirements for all generators, in Order 2003 (issued in 2003) FERC required transmission operators to provide standardized interconnection requirements for all generators, including uniform approaches to placing generators within the interconnection queue.[42] The standardized interconnection process under Order 2003 was to initially proceed as follows:

> First, the prospective Interconnection Customer [generator] will submit an Interconnection Request to the Transmission Provider along with a $10,000 deposit, preliminary site documentation, and the expected In-Service Date [of the generation]. The Transmission Provider will acknowledge receipt of the request and promptly notify the Interconnection Customer if its request is deficient. When the Interconnection Request is complete, the Transmission Provider will place it in its interconnection queue with other pending requests. The Transmission Provider will assign a Queue Position to each completed Interconnection Request based on the date and time of its receipt. Queue Position is used to determine the order of performing the various Interconnection Studies [to test how adding electricity to the line will impact the reliability of the grid] and the assignment of cost responsibility for the construction of facilities necessary to accommodate the Interconnection Request.[43]

Once the request was complete and the generator was placed in the queue, Order 2003 then required the parties to schedule a "scoping meeting" to discuss the points on the grid at which the generator might connect and technical data that would "affect such interconnection options."[44] Next, the operator of the lines to which the generator proposed to interconnect (the Transmission Provider) would either require the generator to prepare studies, or the Transmission Provider itself would prepare studies, to "evaluate the proposed interconnection in detail," determine how the

---

[42] 104 FERC ¶ 61,103 (July 24, 2003).
[43] *Id.* at ¶ 35.
[44] *Id.* at ¶ 36.

interconnection could adversely affect the Transmission Provider's system, and identify changes or additions to the transmission system that would be necessary to "safely and reliably complete the interconnection," including an "Interconnection Feasibility Study," an "Interconnection System Impact Study," and an "Interconnection Facilities Study.[45] Following the study, the generator and Transmission Provider then had to "negotiate the schedule for constructing and completing" upgrades necessary to address any problems identified in the study and incorporate this schedule into the interconnection agreement.[46]

FERC has since changed certain queue procedures to allow some generators to "jump ahead" in the queue—ahead of other generators who submitted earlier interconnection requests—by demonstrating a commitment to building the generation project and actually interconnecting with the particular transmission lines at issue. For example, in 2008 FERC allowed the Bonneville Power Administration, which at one time had approximately 18,000 interconnection requests in its queue,[47] to hold an "open season" for generators in the queue.[48] Generators that wanted their interconnection requests prioritized had a limited time period—the open season—in which to sign a Precedent Transmission Service Agreement, which "committed them to take [transmission] service at a specified time and under specified terms."[49] These generators also had to post a security equal to the cost of one year of transmission service.[50] Generators who did not enter into these agreements were "removed from the queue," at least for the period of the open season, but could re-enter the queue "in future open seasons."[51]

FERC also has allowed MISO to engage in a more nuanced process for managing its queue, in which entities can choose to participate in different phases of the interconnection process or not, and thus move ahead more quickly or not. Generators first enter the queue by submitting an interconnection request and providing certain money and data that show that generators can pass a preliminary interconnection threshold called "milestone 1" or M1.[52] For example, generators must show that they have control over the

---

[45] *Id.*

[46] *Id.* at ¶ 38.

[47] Bonneville Power Admin., Network Open Season Nets Robust Response at 1 (Aug. 2008).

[48] 123 FERC ¶ 61,264 at 9 (June 13, 2008).

[49] Bonneville Power Admin., *supra* note 47, at 1.

[50] *Id.* at 2.

[51] *Id.*

[52] *GI Process Flow Diagram*, MIDCONTINENT INDEP. SYS. OPERATOR, https://www.misoenergy.org/Library/Repository/Study/Generator%20Interconnection/GI%20Process%20Flow%20Diagram.pdf.

site from which they propose to generate electricity or alternatively provide a $100,000 deposit, and they must show that they will not overly impact the stability of the grid.[53] To meet M1's grid stability requirements, wind generators must show, for instance, that they can ramp power up or down by a certain amount if they need to in order to ensure that a steady voltage is maintained in the transmission grid.[54] Generators at this stage also must provide two deposits, which range from $5,000 to $120,000 for each deposit depending on the size of the generator.[55] Generators may then enter the "System Planning and Analysis" phase, during which they complete various studies about the feasibility of interconnecting and decide whether or not they will enter the more serious interconnection phase called the "Definitive Planning Stage," at which point generators can achieve "milestone 2" (M2) by submitting more information and money. At this milestone the required monetary commitments from generators range from $40,000 to $520,000, again depending on generators' size.[56] MISO recently proposed changes to this process[57] and is awaiting FERC technical guidance and approval.[58]

In addition to allowing changes to interconnection queue procedures, FERC also has issued several orders that attempt to standardize interconnection requirements placed on renewable energy generators by transmission providers—the operators of the transmission lines who approve or reject interconnection requests. These orders also have addressed interconnection procedures for all types of small generators.[59] With respect to interconnections of renewable generators, in FERC Order 661 (issued in 2005), FERC attempted to make it easier for wind generators and other renewable energy generators to connect to the grid while also ensuring that these generators, which have intermittent output, did not cause grid reliability issues.[60] The voltage within transmission lines must remain relatively steady to avoid blackouts or brownouts, and sudden surges or reductions of electricity flowing into the grid can cause under- or over-voltages. To prevent these sorts of problems, while also avoiding placing too many obstacles on renewable generators, Order 661 required wind generators to demonstrate a minimum ability to control their power output before connecting to

---

[53] *Id.*
[54] *Id.*
[55] *Id.*
[56] *Id.*
[57] 154 FERC ¶ 61,247 (Mar. 29, 2016).
[58] Amanda Durish Cook, *MISO Queue Changes on Hold Pending Technical Conference*, RTO INSIDER, Apr. 25, 2016.
[59] *See generally* PORTER ET AL., *supra* note 39.
[60] 111 FERC ¶ 61,353 (June 2, 2005).

the grid, and it prohibited transmission line operators from imposing more stringent requirements on these generators before they interconnected. For example, the order provided that if a System Impact Study showed that it was necessary, wind operators would have to "demonstrate the ability to remain online during voltage disturbances"[61] up to a certain time period, meaning the operators would have to show that they could continue providing a minimum amount of electricity for a certain period of time if there was inadequate voltage in the grid. Further, the order required that wind generators have "supervisory control and data acquisition" (SCADA) capabilities, meaning the ability to "transmit data to and receive instructions from the Transmission Provider"[62] in the event that more or less electricity needed to be provided from the wind farm or in other situations in which grid reliability could be compromised.

As shown by the many orders issued by FERC in an effort to open up the electricity grid to more generators, including planning for more transmission lines, approving cost allocation strategies that encourage the construction of new lines, standardizing interconnection procedures, and attempting to speed up queues, improving the grid is not an easy task. But much progress has occurred in recent years. Similar access-based challenges and solutions in the energy transportation sector, including for pipelines, are discussed in Chapter 6. Another challenge faced in all aspects of energy transportation, including for pipelines, is the challenge of ensuring that the energy transportation network is safe and reliable, as discussed in the following section.

III. Safety and Cybersecurity Issues in Energy Transportation

In addition to regulating the siting of energy transportation infrastructure and the cost of this infrastructure, access to it, and the rates that may be charged for it, the federal government and state governments also regulate the reliability and safety of this infrastructure to varying degrees. As threats of physical and cyber-based attacks on this infrastructure have grown,[63] so, too, have

---

[61] *Id.* at ¶ 16, ¶ 26.

[62] *Id.* at ¶¶ 71–72, ¶ 80.

[63] *See, e.g., Energy Delivery Systems Cybersecurity*, DEP'T. OF ENERGY, OFFICE OF ELECTRICITY DELIVERY & ENERGY RELIABILITY http://energy.gov/oe/services/technology-development/energy-delivery-systems-cybersecurity (noting "the critical security challenges of energy delivery systems"); Bill Gertz, *Inside the Ring: U.S. Power Grid Defenseless from Physical and Cyber Attacks*, WASH. TIMES, Apr. 16, 2014 (quoting the president of the North American Electric Reliability Corporation, the public-private federal organization tasked with ensuring grid reliability and security: "I am most concerned about coordinated physical and cyber attacks intended to disable elements of the power grid or deny electricity to specific targets, such as government or business centers, military installations, or other infrastructures."); Siobhan

government efforts to address these safety concerns. This section discusses how governments regulate transmission lines, pipelines, and the transport of oil and ethanol by rail to address safety and security concerns and to ensure that customers receive a steady, reliable supply of the energy product delivered through transportation infrastructure and to prevent accidents associated with this infrastructure.

A. The Electric Grid

Ensuring the reliable delivery of electricity to customers in the quantity in which it is demanded and the time at which it is demanded is an incredibly challenging task. Due to limited availability of electricity storage options, when a customer demands a certain quantity of electricity, that amount must be instantaneously generated and dispatched through the electric grid.[64] Further complicating this situation is the fact that the electricity flowing through the grid must be maintained at a steady voltage—sending too much or too little electricity through the grid can result in under- or over-voltages and brownouts or blackouts,[65] in which electricity stops flowing to customers for a short or longer period. Preventing brownouts and blackouts requires the entity that operates the electric grid to do the following: (1) ensure that generators have committed to supply as much electricity as is needed at a given moment, including during periods of "peak" demand when electricity use spikes;[66] (2) ensure that "ancillary services" are available—services in which generators provide a necessary last-minute infusion of electricity into the grid from reserve power plants or suddenly stop sending electricity through the grid;[67] (3) train utility employees so that when they turn off power at a particular substation, plant, or transmission or distribution line in order to work on equipment, they use adequate protective measures and

---

Gorman, *Electricity Grid in U.S. Penetrated by Spies*, WALL. ST. J., Apr. 8, 2009 (reporting that "[c]yberspies have penetrated the U.S. electrical grid and left behind software programs that could be used to disrupt the system").

[64] V. MAKAROV ET AL., ANALYSIS METHODOLOGY FOR BALANCING AUTHORITY COOPERATION IN HIGH PENETRATION OF VARIABLE GENERATION 1.1 (2010).("Dispatchable generators must be committed and dispatched ahead of time and controlled in real time to follow the load [electricity demanded] variation and ensure an adequate balance of supply and demand.")

[65] *Id.*

[66] *See* Hannah J. Wiseman & Hari M. Osofsky, *Regional Energy Governance and U.S. Carbon Emissions*, 43 ECOLOGY L.Q. 143, 184–87 (2016) (describing how grid operators either rely on individual contracts with utilities, in which utilities commit to providing a certain amount of power when needed, or auctions, in which utilities bid in various amounts of electricity in day-ahead, fifteen-minute, and last-minute auctions, and the grid operator takes the amount of electricity needed to exactly match demand).

[67] *Ancillary Services Market*, PJM LEARNING CENTER, http://learn.pjm.com/three-priorities/buying-and-selling-energy/ancillary-services-market.aspx.

inform other responsible entities connected to the grid so that temporarily shutting down one portion of the grid does not cause other, interconnected portions of the grid to go down;[68] and (4) avoid and quickly respond to physical and cyber incidents that cause the grid to stop functioning properly, such as trees falling on wires, physical attacks on power plants and wires, computer hackers' interfering with the computer software that dispatches electricity from a power plant to the grid, or computer hackers' accessing the software that controls the flow of electricity through different portions of the grid, among other incidents.

The organization responsible for ensuring grid reliability and security in the United States is the North American Electric Reliability Corporation (NERC). For decades, NERC was a self-regulating industry organization. Utilities that owned power plants and transmission lines banded together as part of this organization to agree upon voluntary standards that all members of the organization would follow in order to maintain grid reliability.[69] But after a large blackout in the northeastern United States in 2003, Congress through the Energy Policy Act of 2005 required that NERC be subject to federal oversight. Specifically, Congress required that FERC receive applications from entities that proposed to serve as the "Electric Reliability Organization" (ERO) for the United States,[70] to be overseen and monitored by FERC.[71] NERC was the only entity that applied to be the ERO, and FERC approved NERC as the ERO for the United States in 2006.[72] NERC also has entered into various memoranda of understanding with certain Canadian provinces to serve as the entity that regulates the reliability and security of the electricity system within these provinces.[73]

NERC regulates grid reliability by delegating many of its responsibilities to sub-entities within different regions of the United

---

[68] *Cf.* U.S.–CANADA POWER SYSTEM OUTAGE TASK FORCE, FINAL REPORT ON THE AUGUST 14, 2003 BLACKOUT IN THE UNITED STATES AND CANADA: CAUSES AND RECOMMENDATIONS 18–19 (Apr. 2004) (describing how a lack of human awareness about certain power plants and lines within the grid being down, and thus a failure to dispatch more power and provide more "voltage support," partly contributed to a large blackout in 2003).

[69] N. AM. ELEC. RELIABILITY CORP., HISTORY OF NERC (Aug. 2013), http://www.nerc.com/AboutNERC/Documents/History%20AUG13.pdf.

[70] Energy Policy Act of 2005, Pub. L. 109–58 tit. XII, § 1211 (Aug. 8, 2005) (amending section 215(c) to direct FERC to certify an organization as the ERO if the ERO met certain congressionally-established standards).

[71] *Id.* (amending section 215(b) of the Federal Power Act to provide for FERC oversight of the ERO).

[72] N. AM. ELEC. RELIABILITY CORP., FREQUENTLY ASKED QUESTIONS 2 (Aug. 2013), http://www.nerc.com/AboutNERC/Documents/NERC%20FAQs%20AUG13.pdf.

[73] *Canadian MOUs*, N. AM. ELEC. RELIABILITY CORP., (Aug. 2013), http://www.nerc.com/FilingsOrders/ca/Pages/Canadian-MOUs.aspx

States called "regional entities" (REs).[74] These REs are responsible for proposing reliability standards to NERC, which in turn sends reliability standards to FERC, which rejects or approves the standards or approves them with modifications.[75] Reliability standards are the many rules that power plants and transmission line operators must follow in order to prevent the four types of events mentioned above, and other events, that can cause blackouts and brownouts. Reliability standards include, inter alia, rules requiring utilities and transmission line operators to implement vegetation management programs (including trimming around wires), maintain adequate power plants to serve as "reserves" that can come online during peak periods of electricity demand and when last-minute additional voltage is required, prepare for various emergencies, and take various cyber security measures, among many requirements.[76] REs also assist FERC and NERC in enforcing reliability standards.

Additionally, REs are directly responsible for ensuring that all of their members—power plants and transmission line operators— implement and follow the reliability standards. To do this, REs contract with "reliability coordinators" (RCs) (often a utility) to perform various functions to ensure grid reliability. For example, an RC might train employees to properly isolate a power plant or electrical substation (where high-voltage electricity from transmission lines is reduced to a lower voltage to send through distribution wires) when performing maintenance on this equipment so as to avoid causing electricity voltage reductions throughout the rest of the system.[77] Both REs and RCs can be held directly responsible for a failure to follow reliability standards. For example, in one event in Florida, the Florida Reliability Coordinating Council (FRCC)—the RE for the Florida region—contracted with the utility Florida Power & Light Co. (FPL) to serve as the RC that would implement reliability standard requirements. In February 2008 a field engineer for FPL caused the following problem, which resulted in violations of several reliability standards:

> The event originated at the Flagami Substation on the FPL system when a field engineer was diagnosing a piece of . . . transmission equipment that had previously malfunctioned. In the process, he disabled two levels of

---

[74] *Regional Entities*, N. AM. ELEC. RELIABILITY CORP., http://www.nerc.com/AboutNERC/keyplayers/Pages/Regional-Entities.aspx.

[75] Hari M. Osofsky & Hannah J. Wiseman, *Hybrid Energy Governance*, 2014 U. ILL. L. REV. 1, 36, 41–44.

[76] *United States Mandatory Standards Subject to Enforcement*, N. AM. ELEC. RELIABILITY CORP., http://www.nerc.com/pa/stand/Pages/ReliabilityStandardsUnitedStates.aspx?jurisdiction=United% 20States.

[77] *See* 130 FERC ¶ 61,163 at 1–2, Order Approving Stipulation and Consent Agreement (Mar. 5, 2010).

protection on equipment energized and connected to [the grid] and a "fault" (short circuit) occurred that resulted in transmission outages in the vicinity of the fault as well as generation and distribution outages across portions of the southern two-thirds of the state. . . . The operator fulfilling the RC function was not informed that any protection had been disabled and therefore could not and did not operate the system recognizing [the problems associated with a loss of protection that would have isolated the fault].[78]

As a result of these problems, FERC issued an enforcement order that approved a stipulation and consent agreement between FERC's Office of Enforcement, NERC, and FRCC (the regional entity), in which FRCC "agreed to pay a civil penalty of $350,000" and to "undertake numerous specific reliability enhancement measures."[79]

To address growing physical cyber threats to the grid, FERC has approved several reliability standards that apply specifically to cyber security. For example, in 2014 FERC approved a Physical Security Reliability Standard (CIP–014–1) to address concerns about physical attacks on power plants, transmission lines, and associated infrastructure such as substations.[80] This standard applies to entities that are part of the "Bulk-Power System"[81] (also referred to as the "Bulk Electric System")—in this case, specifically to owners and operators of electric transmission lines. The standard requires, among other things, that these entities identify critical transmission infrastructure that could be vulnerable to physical attack, the types of threats applicable to this infrastructure, and physical security plans and other means of preventing and responding to these threats.[82] These entities must conduct studies and risk assessments in order to properly identify the critical infrastructure and threats, and these documents must be reviewed by independent third parties (with sensitive and confidential information protected from public disclosure).[83]

FERC-approved reliability standards similarly require more members of the Bulk-Power System—including generators, owners and operators of substations, and owners and operators of transmission lines—to protect cyber components of the system.[84]

---

[78] *Id.* at 2.

[79] *Id.* at 1.

[80] 149 FERC ¶ 61,140, Physical Security Reliability Standard (Nov. 20, 2014).

[81] *Id.* at 2–3.

[82] *Id.* at 3, 6–8.

[83] *Id.*

[84] Standard CIP–002–3—Cyber Security—Critical Cyber Asset Identification at 1, http://www.nerc.com/files/CIP-002-3b.pdf.

Specifically, these entities must identify important physical assets of the system, such as transmission lines, power plants, and substations, for which "Critical Cyber Assets" are essential for controlling the physical asset.[85] Examples of Critical Cyber Assets include, inter alia, computers that automatically control power plants (turning them on or off, or ramping them up or down so that they produce more or less electricity), and computers that allow utilities to share generation and power flow data with each other in real-time so that utilities and grid operators know whether the required amount of electricity is flowing through the grid at a given time.[86] Another cyber reliability standard requires that these same entities prepare and implement "a cyber security policy that represents management's commitment and ability to secure its Critical Cyber Assets."[87] As part of this plan, members of the Bulk Power System must maintain the secrecy of certain information about Critical Cyber Assets, such as floor plans for computing centers that operate power flow through transmission lines and control power plants, and operational procedures for these assets.[88] The plan also must list specific "designated personnel" who are allowed to access this protected information, as well as personnel who are allowed to access the Critical Cyber Assets themselves to modify or improve software and conduct other necessary tasks, so that management is aware when a data breach has occurred.[89] Other reliability standards pertaining to cyber security require physical and digital/electronic protection of Critical Cyber Assets, planning for reporting and responding to incidents that cause a compromise of these assets, and training personnel who work with these assets, among other protections.[90]

B. Pipelines

Just as owners and operators of the electric generation and transmission system face a number of challenges from both natural hazards, such as trees, and human-induced threats, such as attacks on physical and cyber assets, pipelines face similar threats. Pipelines that carry oil, natural gas, and other hazardous substances around the country can corrode and leak, be punctured when construction equipment digs underground, experience physical attacks or electronic infiltration of computers that control the flow of substances

---

[85] *Id.* at 2.

[86] *Id.*

[87] Standard CIP–003–3—Cyber Security—Security Management Controls at 1, http://www.ieso.ca/Documents/ircp/2012/NERC/CIP-003-3.pdf.

[88] *Id.* at 2.

[89] *Id.* at 3.

[90] *CIP Standards*, N. AM. ELEC. RELIABILITY CORP., http://www.nerc.com/pa/Stand/Pages/CIPStandards.aspx.

through pipelines, and sometimes explode.[91] The federal agency tasked with addressing the safety and reliability of pipelines and smaller gathering and distribution lines is the Pipeline and Hazardous Materials Safety Administration (PHMSA), which is part of the Department of Transportation. PHMSA's authority comes from the Natural Gas Pipeline Safety Act of 1968[92] (a statute, which, over time, was revised to also include jurisdiction over "hazardous liquid pipeline" facilities[93]), and, most recently, the Pipeline Safety, Regulatory Certainty, and Job Creation Act of 2011, which expands PHMSA authority over natural gas pipelines and gives PHMSA the authority to take enforcement action when oil spills from pipelines occur.[94] Another federal agency, the National Transportation Safety Board (NTSB), is tasked solely with investigating the causes of transportation accidents—including pipeline accidents—and suggesting changes to rules to address risks associated with these accidents.[95] For *intrastate* pipelines, states and municipalities that have received certification from the Department of Transportation may "prescribe and enforce safety standards and practices" for these pipelines.[96]

PHMSA's safety regulations for pipelines include requirements for pipeline design and specifications so that pipelines are constructed of adequately strong material;[97] rules to prevent damage to pipelines as a result of digging and excavation;[98] standards for personnel who manage control rooms for pipelines, including standards for operating during abnormal and emergency conditions and properly changing shifts;[99] maximum allowable operating pressures that vary for pipelines made of different materials;[100] and

---

[91] For a detailed discussion of pipeline and rail safety, *see* Alexandra B. Klass, *Future-proofing Energy Transport Law*, 94 WASH. U. L. REV. ___ (forthcoming 2017), *available at* http://papers.ssrn.com/sol3/Delivery.cfm/SSRN_ID2755991_code702020.pdf?abstractid=2748905&mirid=1.

[92] Pub. L. No. 90–481 (1968) (codified at 49 U.S.C. § 60101–60140, as amended by the Pipeline Safety, Regulatory Certainty, and Job Creation Act of 2011, Pub. L. no. 112–90 (2012)).

[93] Pub. L. no. 103–272 (July 5, 1994) (adding 49 U.S.C. § 60102, providing that "[t]he Secretary shall prescribe minimum safety standards for pipeline transportation and for pipeline facilities" and that such standards "shall be practicable and designed to meet the need for gas pipeline safety, for safely transporting hazardous liquid, and for protecting the environment").

[94] Pub. L. no. 112–90 (2012).

[95] *About the National Transportation Safety Board*, NAT'L TRANSP. SAFETY BD., http://www.ntsb.gov/about/pages/default.aspx.

[96] 49 U.S.C. § 60105 (West 2016).

[97] 49 C.F.R. §§ 192.51–59; 192.101–125 (West 2016).

[98] Pipeline Safety: Pipeline Damage Prevention Programs, 80 Fed. Reg. 43,836 (July 23, 2015).

[99] 49 C.F.R. § 192.631 (West 2016).

[100] 49 C.F.R. §§ 192.619, 192.621, 192.623 (West 2016).

requirements for maintaining, repairing, and monitoring pipelines,[101] among many other standards.

Several dramatic incidents have recently called attention to potentially inadequate standards for pipelines as well as facilities connected to them, including natural gas storage facilities—also regulated by PHMSA—in which natural gas is stored underground and then sent through a pipeline when it is needed. After the explosion of a natural gas pipeline in San Bruno, California, in 2010, "in which 8 people died and more than 50 people were injured,"[102] Congress enacted the Pipeline Safety, Regulatory Certainty, and Job Creation Act of 2011.[103] At that time, PHMSA concluded that "[t]here is a pressing need for an improved strategy to protect the safety and integrity of the nation's pipeline system"[104] in part due to the boom in shale gas production and increased construction and use of pipelines.[105] Responding to recommendations by the NTSB, the Government Accountability Office, and Congress's 2011 Act, PHMSA has recommended that current requirements for assessing the integrity of pipelines and repairing pipelines in high-population areas (called "high-consequence areas") should apply to pipelines in other areas, and it has proposed improved standards for preventing corrosion and excavation damage in "large-diameter, high-pressure" gathering lines that carry gas from wells to pipelines, among other requirements.[106]

Beyond the San Bruno accident, another incident that called attention to PHMSA's regulation of natural gas pipelines and associated infrastructure was the gas leak of a well at the Aliso Canyon Natural Gas Storage facility in 2015, which released approximately 94,500 tons of methane—a potent greenhouse gas—into the air.[107] The leak led to the evacuation of thousands of residents in the nearby Porter Ranch community,[108] and numerous criminal and civil lawsuits have been filed.[109]

As introduced above, natural gas storage facilities are connected to gas pipelines and are used to temporarily inject and store gas

---

[101] 49 C.F.R. §§ 192.703–755 (West 2016).

[102] Pipeline Safety; Safety of Gas Transmission and Gathering Pipelines, 81 Fed. Reg. 20,722, 20,725 (Apr. 8, 2016).

[103] Pub. L. no. 112–90 (2012).

[104] Pipeline Safety, *supra* note 102, at 20,725.

[105] *Id.* at 20,726.

[106] *Id.* at 20,730–31.

[107] *Aliso Canyon Natural Gas Leak*, CAL. ENVTL. PROT. AGENCY AIR RES. BD., http://www.arb.ca.gov/research/aliso_canyon_natural_gas_leak.htm.

[108] Alice Walton, *Assemblyman Mike Gatto Plans Hearings on Porter Ranch Gas Leak*, L.A. TIMES, Dec. 29, 2015.

[109] Paige St. John & Alice Walton, *L.A. County Files Criminal Charges over Porter Ranch Gas Leak*, L.A. TIMES, Feb. 2, 2016.

underground until it is needed. PHMSA regulates any facilities that are connected to interstate pipelines, although states, too, have exerted regulatory authority over these facilities. Indeed, the federal government has left much of the responsibility for the safety of gas storage facilities to the states. In 1997, the Research and Special Programs Administration of the DOT concluded that in light of existing state regulations and industry guidelines, as well as different geology and hydrology in different states, uniform federal natural gas storage safety standards were perhaps not merited.[110] It then encouraged "state action and voluntary industry action as a way to assure underground storage safety instead of proposing additional federal regulations."[111] State regulations vary substantially, however, in terms of maximum injection pressure allowed for gas storage facilities; how the wells into which gas is injected must be constructed; and whether or not the state requires groundwater testing before injection and groundwater monitoring while gas is stored in the facility, among other variations.[112] Further, in at least one case a federal court has held that state underground natural gas storage regulations are preempted by federal law because under the Natural Gas Act, FERC has "exclusive" jurisdiction over interstate transportation of natural gas and "facilities of natural gas companies used in this transportation." According to this court, natural gas facilities connected to interstate pipelines are within FERC's exclusive jurisdiction over "facilities" covered under this provision.[113]

The federal regulations that are in place for natural gas storage facilities are somewhat barebones. FERC regulates natural gas storage facilities primarily on economic grounds; before a natural gas storage facility is constructed, FERC must determine that it is "needed" and must grant a certificate of public convenience and necessity.[114] However, it sometimes attaches environmental conditions to these certificates based on review of the certificate under NEPA.[115] There are few PHMSA safety standards that

---

[110] DEP'T OF TRANSP., Research and Special Programs Administration, Underground Storage of Natural Gas or Hazardous Liquids, Advisory Bulletin (ADB–97–04) (July 10, 1997), http://phmsa.dot.gov/portal/site/PHMSA/menuitem.6f23687cf7b00b0f22e4c6962d9c8789/?vgnextoid=a5d2e0fde382c110VgnVCM1000001ecb7898RCRD&vgnextchannel=8590d95c4d037110VgnVCM1000009ed07898RCRD&vgnextfmt=print.

[111] *Id.*

[112] *See, e.g.*, 14 CAL. CODE REGS. §§ 1724.6, 1724.7, 1724.9, 3403.5, 3227 (West 2016); OKLA. ADMIN. CODE § 165:10–3–5 (West 2016); 52 OKLA. STATS. ANN. §§ 36.1–36.6 (West 2016); 25 PA. CODE §§ 78.401–407 (West 2016).

[113] Colo. Interstate Gas Co. v. Wright, 707 F. Supp.2d 1169, 1174–78 (D. Kan. 2010).

[114] 18 C.F.R. §§ 157.5, 157.6 (West 2016).

[115] *See, e.g.*, Southern Star Central Gas Pipeline, Inc., 115 FERC ¶ 61,219 (May 19, 2006).

specifically apply to underground natural gas storage facilities. This regulatory area is likely to undergo reforms—especially at the state level—in light of the Aliso Canyon incident.

C. Rail

The boom in oil and gas development caused by fracturing has not only increased the use and construction of pipelines but also the transport of oil by rail, as introduced above. Shipments of ethanol, too, also have grown. The PHMSA and the Federal Railroad Administration (FRA) are the two agencies primarily responsible for regulating the rail transport of fuels, and they have recently engaged in an extensive process of issuing emergency orders, recommendations to industries, and updated regulations for rail transport.[116] State and local governments have few opportunities to regulate rail safety. The Federal Railroad Safety Act, from which the FRA obtains much of its authority to govern rail safety, only allows state and local governments to regulate in rail safety areas in which there are no federal orders or regulations "covering the subject matter of the State requirement"[117]—areas that have been interpreted narrowly by the courts.[118] Further, state and local governments may regulate in "covered" rail safety areas only if they can show that the regulation is necessary to address a unique local condition and that the regulation is not incompatible with federal law and avoids unreasonably burdening interstate commerce.[119] Similarly, the Hazardous Materials Safety Act, under which PHMSA regulates, blocks most state regulation, including, for example, regulations that classify certain substances being transported as hazardous or requires these substances to be labeled in a particular way, in addition to many other regulations.[120] If state and local governments manage to find an area not covered by this Act, they may only regulate in a manner that does not conflict with federal law.[121]

---

[116] *See also* Klass, *supra* note 91 (discussing rail safety standards and the responsible agencies); Hannah J. Wiseman, *Negotiated Rulemaking and New Risks: A Rail Safety Case Study*, ___ WAKE FOREST J. L. & POL. ___ (forthcoming 2017) (describing the emergency orders and regulations in detail).

[117] 49 U.S.C § 20106(a)(2) (West 2016).

[118] *See* Hannah J. Wiseman, *Disaggregating Preemption in Energy Law*, 40 HARV. ENVTL. L. REV. 293 (forthcoming 2016) (describing preemption under the Federal Railroad Safety Act and Hazardous Materials Safety Act and how few local or state governments have succeeded in persuading courts that their regulations should not be preempted).

[119] 49 U.S.C. § 20106 (a)(2)(C) (West 2016).

[120] 49 U.S.C. § 5125(b) (West 2016). *See also* Wiseman, *supra* note 118 (describing this preemption).

[121] 49 U.S.C. § 5125(a) (West 2016).

Unlike underground natural gas storage regulation, federal regulation of rail safety is relatively extensive, although not as comprehensive or detailed as some concerned citizens and groups would hope for.[122] In the wake of a growing number of accidents involving derailed oil and ethanol trains,[123] FRSA and PHMSA issued several orders and recommendations to industry in various attempts to improve rail safety,[124] and in the summer of 2015 these agencies issued a new final rule that addresses various aspects of rail safety for trains carrying ethanol and crude oil.[125] This rule creates a category of trains called "high-hazard flammable trains" (HHFTs)[126] and places special regulations on these trains. HHFTs are trains that have "20 or more loaded tank cars" carrying flammable liquid like crude oil or ethanol, with the trains being in a "continuous block" of at least 20 connected cars, or trains that have at least 35 tank cars carrying flammable liquid somewhere within the block of tank cars.[127] The regulations cover the design of tank cars and special braking systems that are required; operational aspects of trains, such as maximum speeds; requirements for better classifying the materials carried in trains; and closer assessment of the routes that HHFTs will follow.[128]

With respect to required technologies and their design, new tank cars that fall within the HHFT definition must meet new standards in terms of design—rather than using the common "DOT–111" tank car, companies moving HHFTs must use "DOT–117" tank cars, which are more resistant to puncture, have stronger fittings over the holes in the tanks where liquids are poured into the tank, have better protection against heat (and are thus less likely to explode or melt), and have other improved designs and materials.[129] Older tank cars must be retrofitted to design and materials standards close to (but not as stringent as) the DOT–117 standard following a retrofit schedule that is ten years for certain types of trains and shorter for

---

[122] *See* Hazardous Materials: Enhanced Tank Car Standards and Operational Controls for High-Hazard Flammable Trains, 80 Fed. Reg. 26,644 (May 8, 2015) (describing citizens' and environmental groups' comments on the most recent final rule addressing the safety of oil and ethanol trains).

[123] *Id.* at 26,645 (noting that "in recent years, train accidents/incidents (train accidents) involving the release of a flammable liquid and resulting in fires and other severe consequences have occurred").

[124] *Chronology of DOT Actions on Safe Transportation of Flammable Liquids by Rail*, U.S. DEP'T OF TRANSP., https://www.transportation.gov/mission/safety/rail-chronology.

[125] Hazardous Materials, *supra* note 122, at 26,644.

[126] *Id.*

[127] *Id.* at 26,645.

[128] *Id.* at 26,646 (summarizing these requirements).

[129] *Id.* at 26,666.

other types.[130] All HFFTs with 70 or more loaded tank cars that travel at speeds greater than 30 miles per hour must have special electronically-controlled braking systems.[131]

With respect to operating these redesigned trains, HHFTs may travel at a maximum of 50 miles per hour, and a maximum of 40 miles per hour in "High Threat Urban Areas" (HTUAs), which are areas with "one or more cities," including a 10-mile buffer area around the city or cities.[132]

Maintaining the safety and reliability of the infrastructure used to transport all types of fuels, whether this infrastructure involves a transmission line, pipeline, or rail car, is a complex task. As the United States has expanded the production of fuels in recent years and experienced more threats relating to physical and cyber attacks, this Chapter has shown that Congress and agencies have, to varying degrees, changed statutes and regulations and issued new ones in order to respond to new and growing risks. Further changes to safety and reliability laws are likely to continue in the near future, particularly in light of the fact that much of this infrastructure is aging.

---

[130] *Id.* at 26,682–83.
[131] *Id.* at 26,646, 26,696.
[132] *Id.* at 26,646.

Chapter 4

ENVIRONMENTAL REGULATION OF ENERGY PRODUCTION AND USE

I. Introduction

A wide range of environmental laws and regulations apply to the production, transportation, use, and disposal of energy resources. Nearly all of the major federal environmental statutes enacted since 1969 apply directly to or otherwise affect the energy industry. Moreover, most states and many local governments have their own environmental protection laws that impact energy production, transportation, use, and disposal. This Chapter does not purport to discuss all of the environmental laws that may apply in the energy area. Instead, it focuses more narrowly on the federal and state environmental laws that apply to a discrete set of energy activities to illustrate the range of federal and state regulation of energy. The first section of this Chapter focuses on the environmental regulation of underground and surface coal mining, with an emphasis on the Surface Mining Control and Reclamation Act (SMCRA). The second section turns to federal, state, regional, and local environmental regulation of oil and gas development, including hydraulic fracturing—a topic discussed in more detail in Chapter 2. Next, the third section turns to federal and state efforts to reduce greenhouse gas (GHG) emissions and other air pollutants from the electric power sector, particularly from coal-fired power plants. This section includes a discussion of the federal Clean Power Plan, the northeastern states' Regional Greenhouse Gas Initiative (and other regional initiatives), and state laws including California's Global Warming Solutions Act and state renewable portfolio standards (RPSs). The fourth section then discusses the regulation and licensing of hydropower facilities, with a focus on the environmental protection aspects of that process. Finally, the last section discusses environmental issues surrounding the licensing of nuclear power plants and nuclear waste disposal.

II. Environmental Regulation of Coal Extraction

Approximately half of U.S. coal production comes from underground mines (higher Btu bituminous coal), mostly in the eastern half of the country, and the other half is produced from surface mines with large coal seams (subbituminous coal) in western states, primarily in the Powder River Basin of Wyoming and Montana. This section discusses the primary environmental, health, and safety laws governing underground mining and surface mining.

The Mine Safety and Health Act of 1977 (Mine Act)[1] is the primary statute governing federal health and safety regulation of the mining industry. The Mine Act amended the 1969 Coal Act and transferred jurisdiction over health and safety in mines from the U.S. Department of the Interior (DOI) to a new agency within the Department of Labor—the Mine Safety and Health Administration (MSHA).[2] MSHA's purpose is to "promote safety and health in the mining industry" and "prevent recurring disasters" in this industry."[3] MSHA has jurisdiction over all mines in the United States, which includes conducting inspections (four annual inspections of each underground mine and two annual inspections of each surface mine[4]) and for investigating accidents and issuing citations for regulatory violations.[5] As explained by the Department of Labor, the "Mine Act also established the independent Federal Mine Safety and Health Review Commission to provide for independent review of MSHA's enforcement actions."[6] Further statutory changes emerged in 2006, when Congress enacted the Mine Improvement and New Emergency Response Act (MINER Act).[7] The MINER Act amended the Mine Act in an effort to enhance safety, requiring each operator to "develop and adopt a written accident response plan" for emergencies;[8] limit liability for individuals and employers involved in mine accident rescue operations;[9] mandate new regulation of mine rescue teams;[10] require notification of the Secretary of Labor within 15 minutes of an incident;[11] and increase the penalties associated with citations as well as willful violations of health and safety standards,[12] among other provisions. While all mining operations pose risks to human health and safety, underground mines in particular create significant risks of accidents, death, and subsidence and well as the risk of "black lung disease," a serious lung disease caused by long-term inhalation of coal dust.

As for surface mining, which poses different health, safety, and environmental concerns than underground mining, the primary

---

[1] 30 U.S.C. §§ 801, *et seq.* (West 2016).

[2] *History of Mine Safety and Health Legislation*, U.S. DEP'T OF LABOR, MSHA, http://arlweb.msha.gov/MSHAINFO/MSHAINF2.htm.

[3] Pub. L. no. 95–164, 91 Stat. 1290 (Nov. 9, 1977).

[4] 30 U.S.C. § 813(a) (West 2016).

[5] *See History of Mine Safety and Health Legislation*, U.S. DEP'T OF LABOR, MSHA, *supra* note 2.

[6] *Id.*

[7] *Id.*; Pub. L. no. 109–236, 120 Stat. 493 (2006).

[8] Pub. L. no. 109–236, § 2 (codified at 30 U.S.C. § 876).

[9] *Id.* at § 3 (codified at 30 U.S.C. § 826).

[10] *Id.* at § 4 (codified at 30 U.S.C. § 825).

[11] *Id.* at § 5 (codified at 30 U.S.C. § 813).

[12] *Id.* at § 8 (codified at 30 U.S.C. § 820).

environmental law governing coal extraction is the Surface Mining Control and Reclamation Act of 1977 (SMCRA).[13] There are several different types of surface mining where coal is removed from surface outcrops or where the overburden (over-lying rock) is removed to extract the coal: (1) contour mining, which involves surface mining along sloping terrain; (2) area mining where a large "box" of overburden is removed to extract the coal, and the overburden is then placed in the previous "box" to remediate the area; and (3) mountaintop mining, which is an adaptation of area mining to mountainous areas, and is most common in the Appalachian states. With mountaintop mining, the overburden is removed from the mountaintop (often taking off 200–600 feet of mountains and ridges), and the coal seams are mined. While some of the overburden is left on the mountaintop, much of the overburden is deposited in neighboring valleys, filling in creeks and waterways.

As noted above, approximately half of U.S. coal comes from surface mines, and much of that underlies federal lands. Until the early 1900s, federal land and mineral policy was to sell certain lands in the public domain outright in fee simple.[14] However, several statutes enacted in the homesteading era of the early 1900s required the United States to retain coal and other mineral resources, conveying only the surface estate to private parties.[15] In 1920, Congress enacted the Mineral Leasing Act (MLA)[16] to govern the private exploration, development, and removal of coal and other fuel minerals on federal lands. The MLA authorizes the Secretary of the Interior, through the Bureau of Land Management (BLM), to lease coal and other subsurface mineral rights for development.[17] Congress has amended the MLA through the Federal Coal Leasing Amendments Act of 1976 to address, among other things, competitive leasing, diligent development of mineral resources, and BLM's receipt of fair market value of use of public resources.[18] In 2016, President Obama placed a moratorium on new coal leases on federal lands in order to re-examine the federal coal-leasing program's impacts on health and the environment, and also to re-evaluate the

---

[13] 30 U.S.C. § 814 (West 2016).

[14] *Powder River Basin Coal, History of the Coal Program*, U.S. DEP'T OF THE INTERIOR, BUREAU OF LAND MGMT., http://www.blm.gov/wy/st/en/programs/energy/Coal_Resources/PRB_Coal/history.html.

[15] *Id.*

[16] 30 U.S.C. §§ 181, *et seq.* (West 2016).

[17] 30 U.S.C. § 181 (West 2016).

[18] U.S. DEP'T OF THE INTERIOR, BUREAU OF LAND MGMT., *supra* note 14.

royalty rates companies must pay to the federal government for coal extraction from federal lands.[19]

Until the enactment of SMCRA in 1977, states had primary authority over coal mining on state and private lands. In SMCRA, however, Congress created the Office of Surface Mining Reclamation and Enforcement (OSM) as a new agency within the DOI.[20] OSM has primary responsibility for administering the new programs created in SMCRA that regulate surface coal mining operations and the surface impacts of underground coal mining operations. Under SMCRA, anyone wishing to conduct a new surface coal mining operation or re-open an old surface mine, or to mine over or through an underground mine, must first obtain a permit from a state with approved authority to implement SMCRA, or from the federal government.[21]

Congress enacted SMCRA to "establish a nationwide program to protect society and the environment from the adverse effects of surface coal mining operations" while assuring an adequate supply of coal to meet U.S. energy needs.[22] But even though SMCRA created nationwide standards, it provided for significant state involvement by allowing states to adopt their own programs to implement the law—specifically, by permitting states to "assume exclusive jurisdiction over the regulation of surface coal mining and reclamation operations," with federal approval (and some exceptions in which the federal government retains authority).[23] Once the Secretary of the Interior approves a state program, the state is the primary regulator of surface mining on non-federal lands, and most coal-mining states now have delegated authority from OSM to regulate coal mining on state lands.[24] Of the coal producing states and tribal lands, 23 have primacy and only one state with active coal mining—Tennessee—has a federally administered program.[25] This is somewhat different than the traditional "cooperative federalism" model of other environmental legislation of that era such as the Clean Air Act (CAA) and the Clean Water Act (CWA), where the U.S. Environmental Protection Agency (EPA) retains significantly more

---

[19] Joby Warrick & Juliet Eilperin, *Obama Announces Moratorium on New Federal Coal Leases*, WASH. POST, Jan. 15, 2016, https://www.washingtonpost.com/news/energy-environment/wp/2016/01/14/obama-administration-set-to-announce-moratorium-on-some-new-federal-coal-leases/.

[20] Pub. L. no. 95–87 Tit. II, § 201, 91 Stat. 445 (Aug. 3, 1977) (codified at 30 U.S.C. § 1211).

[21] *Id.* at Tit. V, §§ 502, 503 (30 U.S.C. §§ 1252, 1253).

[22] *See* 30 U.S.C. § 1202(a) and (f) (West 2016).

[23] Pub. L. no. 85–87 Tit. V § 503 (codified at 30 U.S.C. § 1253).

[24] *Regulating Coal Mines*, OFFICE OF SURFACE MINING RECLAMATION AND ENF'T, http://www.osmre.gov/programs/rcm.shtm.

[25] *Id.* (Iowa also has primacy but does not have active coal mining.)

authority to veto permits or otherwise exert influence even where the state has delegated authority. By contrast, under SMCRA, once a state has "primacy," its regulatory authority over surface mines is virtually exclusive.[26] However, SMCRA still retains OSM oversight authority over states with primacy to ensure compliance with federal standards by empowering the Secretary of the Interior, through OSM, to conduct investigations and inspections of the state program and to take federal enforcement action if the state program is determined to not be effectively enforcing SMCRA's provisions. OSM also augments state efforts. As the Office explains, "The OSM also partners with states and Indian tribes to regulate mining on federal lands and to support States' regulatory programs."[27]

SMCRA is made up of two main parts. Title IV—"Abandoned Mine Reclamation," creates the "Abandoned Mine Reclamation Fund," funded by a reclamation fee that is charged per ton of coal produced from surface or underground operations.[28] The fund is to be used for "reclamation and restoration of land and water resources adversely affected by past coal mining" and for filling "voids and sealing of tunnels, shafts, and entryways" in old mines, among other purposes.[29] Title V of the Act provides environmental standards for surface coal mining, "surface effects of underground coal mining operations," and disposal of wastes from these mining operations.[30] Under this title, operators proposing to conduct surface mining or underground mining with surface effects must apply to the state (if the state has primacy) or OSM for a permit by submitting a mining application and reclamation plan.[31] The application must include information such as a description of the method of proposed mining, how the mining operation is likely to affect the hydrology in the area, and the locations of surface waters where mining wastes will be disposed of.[32] The relevant authority may only issue a permit if the operator shows that it will be able to accomplish the reclamation required by the Act; the operation is designed to prevent "material damage" to the overall hydrologic balance (the flow of water aboveground and underground) beyond the area being mined; the area of proposed mining has not been designated as unsuitable for mining under SMCRA; and the operation will "not interrupt, discontinue, or preclude farming" in river valleys and "not materially

---

[26] *See, e.g.,* Bragg v. W. Va. Coal Ass'n, 248 F.3d 275 (4th Cir. 2001).

[27] *See Regulating Coal Mines,* OFFICE OF SURFACE MINING RECLAMATION AND ENF'T, May 21, 2015, http://www.osmre.gov/programs/rcm.shtm.

[28] Pub. L. no. 95–87 Tit. IV, §§ 401, 402 (30 U.S.C. §§ 1231, 1232)).

[29] 30 U.S.C. § 1231 (West 2016).

[30] Pub. L. no. 95–87 Tit. V.

[31] *Id.* at Tit. V, § 510 (codified at 30 U.S.C. § 1260).

[32] *Id.* at Tit. V, § 515 (codified at 30 U.S.C. § 1265).

damage the quantity or quality of water in surface or underground water systems that supply these valley floors."[33] SMCRA also requires operators to post a bond prior to commencing mining operations to ensure that money will be available for the government to reclaim the mining site once mining operations are complete in the event that the operator fails to do so.[34] Further, before commencing mining, the operator must obtain permits under the CWA if the operator plans to dispose of mining wastes in wet areas, which is a common method of disposal.

After the completion of mining operations, SMCRA requires the operator of most types of mines to restore or "reclaim" the affected land to a condition capable of supporting the uses it could support before mining, or to "higher or better uses."[35] The operator must also: (1) "restore the approximate original contour ['AOC'] of the land" and, after restoration, "contour, backfill, grade, and compact (where advisable) the excess overburden and spoil"; (2) "minimize ... disturbances to the prevailing hydrologic system" and beyond the mine site by "avoiding acid or other toxic mine drainage" through measures such as preventing water from touching toxic deposits and treating drainage; (3) on agricultural lands, separate out and preserve topsoils and "replace and regrade the root zone material"; and (4) in regraded areas, "establish ... a diverse, effective, and permanent vegetative cover" similar to the variety of vegetation previously on the site."[36] Some of these requirements do not apply to mountaintop removal mining, for which the land could not feasibly be restored to its AOC.

As noted earlier, one form of surface mining (and a particularly controversial one) is known as "mountaintop mining." Mountaintop mining involves removing the top of a mountain in order to recover the coal seams contained within it. The practice is widespread in six Appalachian states (Kentucky, West Virginia, Virginia, Tennessee, Pennsylvania, and Ohio). It creates large quantities of excess dirt or "spoil" which is generally disposed of in valley fills on the sides of the former mountain, burying streams that flow through the valley.

Both SMCRA and the CWA § 404 permit process govern the deposit of overburden from mountaintop mining. CWA § 404 requires a permit for the deposit of "dredge and fill" material into wetlands.[37] While the CWA prohibits "discharges" of "pollutants" (which include dirt) into navigable waters or adjacent wetlands without a permit,

---

[33] *Id.* at Tit. V, § 510 (codified at 30 U.S.C. § 1260).
[34] *Id.* at Tit. V, § 509 (codified at 30 U.S.C. § 1259).
[35] 30 U.S.C. § 1265 (West 2016).
[36] *Id.*
[37] 33 U.S.C. § 1304 (West 2016).

the U.S. Army Corps of Engineers, which has permitting authority under CWA § 404 with EPA oversight and veto authority, has traditionally permitted the deposit of fill from surface coal mining activities into waters of the United States under Nationwide Permit 21 (NWP 21). There has been a significant amount of additional litigation over whether NWP 21 as applied in this context is consistent with the CWA. This is because nationwide permits, which are designed for an entire class of activities without separate environmental review of the particular project, are only supposed to be issued for activities that have minimal environmental effects. While the Army Corps defended its use of NWP 21 during the Bush II administration, the Army Corps then changed its approach. As explained by the Congressional Research Service, in June 2010, the Army Corps under the Obama Administration suspended the use of NWP 21 in the Appalachian region "while continuing to evaluate modification or suspension of the permit."[38] In March 2012, the Army Corps reissued NWP 21 with limits for the acres and feet of streambeds that could be destroyed, and it prevented the Army Corps from using the nationwide permit process to allow "valley fills," in which the overburden from surface mining is dumped within the headwaters of a stream.[39] Because this change substantially limited mining companies' ability to seek approval for disposing of mining waste under NWP 21, the Army Corps provided an alternative review process, requiring the companies to apply for a section 404 permit under the CWA, which involves more significant environmental review of the specific project.[40]

The EPA also has been more active in reviewing Army Corps dredge and fill permits for the dumping of overburden in streams. The chronology of events, as summarized by the Congressional Research Service, is as follows. In 2011, the EPA retroactively vetoed a CWA § 404 permit the Army Corps had issued to the large Spruce I mine project in West Virginia in 2007. The waste from the mine would have buried "over seven miles of streams," impacted "2,278 acres of forestland," and degraded "water quality in streams adjacent to the mine."[41] After the mine owners filed suit in federal court in 2013, the U.S. Court of Appeals for the D.C. Circuit upheld EPA's authority to veto the permit based on concerns about environmental impacts, holding that the plain language in section 404(c) grants EPA

---

[38] CLAUDIA COPELAND, CONG. RESEARCH SERV., MOUNTAINTOP MINING: BACKGROUND ON CURRENT CONTROVERSIES 11 (Apr. 20, 2015), https://www.fas.org/sgp/crs/misc/RS21421.pdf.

[39] Id.

[40] Id.

[41] COPELAND, supra note 38, at 14.

authority to veto Army Corps wetlands permits even after the permit has been issued. The Court reasoned:

> Section 404 imposes no temporal limit on the Administrator's authority to withdraw the Corps's specification but instead expressly empowers him to prohibit, restrict or withdraw the specification *"whenever"* he makes a determination that the statutory "unacceptable adverse effect" will result.[42]

The OSM has also engaged in new rulemaking to address mountaintop mining. In June 2009, the EPA, Army Corps and OSM signed a Memorandum of Understanding implementing an interagency plan designed to "significantly reduce the harmful environmental consequences of Appalachian surface coal mining operations."[43] One of the actions the MOU addressed was a revision of the OSM's Stream Buffer Rule, last amended in 2008.[44] This rule, as summarized by the Congressional Research Service, is designed to reduce the impacts of coal mining on "surface water, groundwater, fish, wildlife, and other natural resources by limiting the placement of [coal] waste in streams and limiting the generation of mining waste."[45] In July 2015, OSM proposed a new Stream Protection Rule that would revise the regulations that implement Title V of SMCRA.[46] The proposed rule is designed to "collect adequate premining data about the site of the proposed mining operation" so that regulatory officials have adequate information on baseline conditions and can better assess the impacts of mining; "adjust monitoring requirements to enable timely detection and correction of any adverse trends" in water quality or quantity or in stream biological conditions; better protect and restore streams; and "ensure that land disturbed by mining operators is restored" to a point where it is able to support premining use, among other goals.[47]

---

[42] Mingo Logan Coal Co. v. EPA, 714 F.3d 608, 613 (D.C. Cir. 2013) (quoting 33 U.S.C. § 1344(c) (emphasis added)).

[43] Memorandum of Understanding Among the U.S. Department of the Army, U.S. Department of the Interior, and U.S. Environmental Protection Agency, Implementing the Interagency Action Plan on Appalachian Surface Coal Mining (June 11, 2009), http://www.osmre.gov/resources/mou/ASCM061109.pdf.

[44] *Building a Stream Protection Rule*, OFFICE OF SURFACE MINING RECLAMATION AND ENF'T, http://www.osmre.gov/programs/rcm/streamprotectionrule.shtm.

[45] CLAUDIA COPELAND, CONG. RESEARCH SERV., THE OFFICE OF SURFACE MINING'S PROPOSED STREAM PROTECTION RULE: AN OVERVIEW (Sept. 10, 2015), http://www.osmre.gov/resources/reports/CRS_StreamProtectionRule.pdf.

[46] *Id.*

[47] Dep't of the Interior, Office of Surface Mining Reclamation & Enf't, Stream Protection Rule (Proposed Rule), http://www.osmre.gov/programs/RCM/docs/SPRProposedRule.pdf.

III. Environmental Regulation of Oil and Gas Operations

Oil and gas operations have had and continue to have adverse impacts on environmental resources such as air, water, and land. These adverse impacts include, among others: (1) air pollution emissions from drilling and hydraulic fracturing equipment, the hydraulic fracturing process, hydrocarbon storage tanks, natural gas flaring, and emissions from vehicles that drive to and from sites as well as engines that run on oil and gas sites; (2) water pollution from wastewater produced during the drilling and fracturing process, improper disposal of wastes to surface and groundwater, and spills and leaks from pipelines, gathering lines, wellheads, storage tanks, pits, and transport trucks; (3) impacts to aquatic life when too much water is withdrawn from a surface water source at a given time, as well as other water use impacts from groundwater and surface water withdrawals; and (4) impacts on wildlife and plant life as a result of water withdrawals, site development and associated habitat fragmentation and disruption, and bird deaths caused by open wastewater pits.[48]

As discussed in Chapter 2, apart from oil and gas leasing on federal lands, state law has governed much of onshore oil and gas development since its inception. Nevertheless environmental protection and worker safety laws, including, inter alia, the CAA,[49] Oil Pollution Act (OPA), Endangered Species Act, and Occupational Safety and Health Act now place certain regulatory restrictions or reclamation requirements on such development. CWA or Safe Drinking Water Act (SDWA) restrictions also apply to the disposal of oil and gas wastes depending on the method of disposal,[50] and oil and gas operators are liable for cleaning up sites contaminated by releases of hazardous substances other than oil and gas under the

---

[48] *See* Margriat F. Carswell, *Balancing Energy and the Environment*, in THE ENVIRONMENT OF OIL 179, 182–85 (Richard J. Gilbert, ed. 1993); Hannah J. Wiseman, *Risk and Response in Fracturing Policy*, 84 U. COLO. L. REV. 729 (2013).

[49] The EPA has finalized rules under the CAA that apply to all newly hydraulically fractured and refractured gas wells and limit volatile organic compound (VOC) emissions from these wells during fracturing. 40 C.F.R. § 60.5375(a) (1)-(4) (West 2016). Additionally, the EPA in 2016 issued a final rule that limits methane emissions from newly hydraulically fractured and refractured oil wells. Oil and Natural Gas Sector: Emission Standards for New, Reconstructed, and Modified Sources, 81 Fed. Reg. 35,824 (June 3, 2016).

[50] The CWA limits the disposal of salty water that naturally comes out of oil and gas wells along with oil and gas. West of the 98th U.S. Meridian, this wastewater may be disposed of in surface waters after being treated for oil and grease. 40 C.F.R. §§ 435.50, 435.52 (West 2016). East of the 98th Meridian it may not be disposed of in surface waters. 40 C.F.R. § 435.32 (West 2016). All oil and gas wastes disposed of through injection into an underground disposal well require a Safe Drinking Water Act permit for the disposal well. 42 U.S.C. § 300h (West 2016).

Comprehensive Environmental Response, Compensation, and Liability Act (CERCLA). Moreover, for oil and gas development on federal lands, the Federal Land Policy and Management Act (FLPMA), directs the Secretary of the Interior to ensure that public lands are managed "under principles of multiple use and sustained yield,"[51] and to "take any action necessary to prevent unnecessary or undue degradation of the [public] lands."[52] In 2015 the BLM issued regulations for hydraulic fracturing on federal lands using its FLPMA and MLA authority, but a federal district court stayed the enforcement of these regulations[53] and then invalidated the rule on the merits.[54] The BLM, Sierra Club, and other parties have appealed both the injunction and the merits decisions. As in many other areas of energy law, the application of these federal environmental laws to specific forms of energy production attempt to balance the desire for cost-effective energy development with environmental protection goals.

Beyond the balancing of energy development and environmental protection required under federal laws like FLMPA, federal environmental laws contain significant exemptions for oil and gas operations. One of the broadest exemptions of oil and gas development from state law is the exemption of most oil and gas wastes from the hazardous waste portion of the Resource Conservation and Recovery Act (RCRA), which regulates the generation, transport, and disposal of wastes.[55] This means that oil and gas operators need not comply with federal requirements when holding most wastes on site in tanks or pits or disposing of wastes on- or offsite, even if the wastes are hazardous.

A second broad-based exemption for oil and gas wells is an exemption from the CAA. Under the CAA, stationary sources of air pollutants are not subjected to the most rigorous CAA requirements unless their potential to emit pollutants exceeds a certain minimum threshold. In many cases, oil and gas wells are not considered a "major" source of air pollution—and thus are not subjected to the most rigorous CAA requirements—unless they are aggregated with nearby pipelines and compressor stations that compress gas in order to send it efficiently through the pipeline. When considering the emissions of wells and compressor stations together, the source usually counts as "major." However, the CAA prohibits this type of

---

[51] 43 U.S.C. § 1732(a) (West 2016).

[52] 43 U.S.C. § 1732(b) (West 2016).

[53] Wyoming v. U.S. Dep't of the Interior, 136 F.Supp.3d 1317 (D. Wyo. 2015).

[54] Wyoming v. U.S. Dep't of the Interior, 2016 WL 3509415 (D. Wyo., June 21, 2016).

[55] Regulatory Determination for Oil and Gas and Geothermal Exploration, Development and Production Wastes, 53 Fed. Reg. 25,446, 25,456 (July 6, 1988).

aggregation with respect to regulating hazardous air pollutants from oil and gas wells.[56]

Another federal exemption applies specifically to hydraulic fracturing. In the Energy Policy Act of 2005, Congress amended the SDWA—which requires entities injecting substances underground to ensure that they will not "endanger" underground sources of drinking water—to exclude hydraulic fracturing, with the exception of fracturing that uses diesel, from the definition of "injection."[57] Thus, oil and gas operators that conduct a non-diesel fracturing operation need not first obtain a federal SDWA permit or ensure that they will avoid the endangerment of underground drinking water. This exemption is sometimes described as the "Halliburton loophole" because companies that conduct hydraulic fracturing around the country, including Halliburton, lobbied for the exemption.

A final exemption for oil and gas operators is from CERCLA. Although oil and gas operators are liable for the cleanup and restoration of sites where these operators have spilled hazardous fracturing chemicals and other hazardous substances, they are not liable for spills of oil and gas.[58] The OPA, however, does trigger certain liability for these types of releases.

Thus, at the federal level several environmental statutes and regulations that implement these statutes apply to oil and gas drilling and fracturing activity, whereas oil and gas activities are largely exempted from other important environmental statutes and regulations.

Although federal, state, and local governments enact and enforce the bulk of environmental regulations governing energy production, use, and disposal, there are also examples of "regional governance" of environmental resources, which can impact oil and gas development. The primary example of this is the Delaware River Basin Commission (DRBC).

The DRBC was formed by a "compact" (agreement) among its member states—New York, New Jersey, Pennsylvania, and Delaware—and the United States.[59] Whenever states collaboratively regulate cross-border issues, such as water quantity and quality, they must obtain congressional approval under the Compact Clause of the U.S. Constitution because they would otherwise impermissibly interfere with interstate commerce. Thus, when the states that are

---

[56] 42 U.S.C. § 7412(n)(4) (West 2016).
[57] 42 U.S.C. § 300h(d)(1) (West 2016).
[58] 42 U.S.C. § 9601(14) (West 2016).
[59] DEL. RIVER BASIN COMM'N, DELAWARE RIVER BASIN COMPACT (1961, reprinted 2009), http://nj.gov/drbc/library/documents/compact.pdf.

now members of the DRBC determined that shared issues associated with the river posed too much of a threat (including the threat of ongoing litigation among the states), they formed the Commission in order to address a variety of issues, from permitting water withdrawals to regulating activities that impact water quality. The Commission is comprised of the Governors of the four member states and a non-voting Army Corps of Engineers Representative.[60] It has a staff of about 30 members, including hydrologists and water resource engineers.[61]

A large portion of the Delaware River Basin overlaps with the Marcellus Shale,[62] and it has become an area of significant hydraulic fracturing activity since approximately 2006. In May of 2009, the DRBC executive director issued a determination that all natural gas production activity in the basin needed to be reviewed by the Commission.[63] The DRBC subsequently worked to draft regulations that would address the impacts of drilling within the basin—proposing not just to regulate water withdrawals within the basin but also to regulate runoff from well sites and other activities that could affect water quality. As the Commission worked to draft the regulations it placed a moratorium on drilling in certain counties within the watershed.[64] After issuing draft regulations the State of New York sued the Army Corps of Engineers (a member of the DRBC) in federal court, arguing that the regulations required review under the National Environmental Policy Act due to their environmental impacts. A federal district court held that these litigants lacked standing because the rules were still in draft form, and there had not yet been an injury-in-fact,[65] but state opposition did not end with this case. After the DRBC issued revised draft rules, Governor Jack Markell of Delaware announced that he would oppose the rules,[66] and the rules still have not been finalized.

The Susquehanna River Basin Commission (SRBC)—another regional entity approved around the same time as the DRBC—allows

---

[60] *About DRBC*, DEL. RIVER BASIN COMM'N, http://www.nj.gov/drbc/about/.

[61] *Id.*

[62] State Impact, *Delaware River Basin Commission: Battleground for Gas Drilling*, NAT'L PUB. RADIO, https://stateimpact.npr.org/pennsylvania/tag/drbc/.

[63] *May 2009—DRBC Eliminates Review Thresholds for Gas Extraction Projects in Shale Formations in Delaware Basin's Special Protection Waters*, DEL. RIVER BASIN COMM'N, http://www.state.nj.us/drbc/home/newsroom/news/approved/20090519_news rel_naturalgas.html.

[64] State Impact, *supra* note 62.

[65] New York v. U.S. Army Corps of Eng'rs, 896 F. Supp.2d 180, 195 (E.D.N.Y. 2012).

[66] Gov. Jack A. Markell, *Markell: "Fracking" Proposal Currently Lacks Sufficient Health and Safety Protections*, DELAWARE.GOV, Nov. 17, 2011, http://news.delaware.gov/2011/11/17/drbc_fracking/; *see also* State Impact, *supra* note 62 (describing this course of events).

hydraulic fracturing within its watershed but regulates withdrawals of water for fracturing. Specifically, any oil and gas operator wishing to withdraw water for fracturing must apply to the SRBC for a permit and indicate the proposed location of the withdrawal and how much water will be withdrawn at a particular rate. For proposed surface water withdrawals, when the Commission rejects or approves the withdrawal it considers how the water withdrawal will impact the minimum amount of water that will be in a stream, creek, or surface water at a given time, and how the withdrawal will impact minimum flow and thus potentially impact aquatic species.[67]

Beyond regional governance of oil and gas, state and local governments play an important role in regulating the impacts of oil and gas development. As discussed in more detail in Chapter 2, the extent to which local governments may regulate oil and gas development varies among states. But where local governments are allowed to impose restrictions on this development, many of them establish zoning districts in which oil and gas development is or is not permitted (or is allowed conditionally and requires case-by-case approval), require environmental liability insurance and other insurance protections for operators in the event that something goes wrong at a well, and impose noise-based and aesthetic restrictions, among others. States, in turn, are primarily responsible for permitting oil and gas wells, requiring protective equipment to be used at well sites to prevent "blowouts" (explosions caused by an uncontrolled build-up of pressure); mandating specific types of and depths of casing (steel lining) to be cemented into the well to prevent oil and gas from leaking into groundwater; regulating the storage and disposal of oil and gas wastes; and inspecting well sites. States' responsibilities for regulating oil and gas development are also enhanced by the fact that the EPA has delegated to the states many federal responsibilities for permitting, such as permitting disposal wells into which oil and gas wastes are injected.

One issue that has arisen with state-issued permits for oil and gas wells and oil and gas waste disposal wells is whether an administrative permit displaces a common law claim—in other words, whether a permitted well that pollutes water or causes another problem is protected from tort claims and other common law claims. So far, state courts have typically answered this question in the negative. For example, in Oklahoma, plaintiffs sued companies that owned and operated oil and gas wastewater disposal wells, arguing that these wells caused earthquakes that injured the

---

[67] *Frequently Asked Questions (FAQs), SRBC's Role in Regulating Natural Gas Development*, SUSQUEHANNA RIVER BASIN COMM'N, http://www.srbc.net/programs/natural_gas_development_faq.htm.

plaintiffs. The Oklahoma Supreme Court found that the permit approving the disposal well—issued by Oklahoma's Corporation Commission—did not preclude a common law claim. Indeed, the court noted that the Corporation Commission's "jurisdiction is limited solely to the resolution of public rights" and that only a court could properly hear a private legal dispute between individuals and the companies that operated the wells.[68] Similarly, in a Texas case that is applicable to oil and gas waste disposal wells, a disposal well that contained non-oil and gas wastes appeared to have leaked pollution, and the pollution appeared to have entered neighboring property. The Texas Supreme Court found that "[a]s a general rule, a permit granted by an agency does not act to immunize the permit holder from civil tort liability from private parties for actions arising out of the use of the permit."[69] State regulation of oil and gas development is further discussed in Chapter 2.

IV. Reducing Air Pollution from the Transportation Sector: Vehicle Emissions Standards and Biofuels Policies

In contrast with the environmental regulation of oil and gas, which rests primarily at the state level, regulation of the impacts of transportation—cars, trains, airplanes, and similar modes of transport—is largely federal. The transportation sector constitutes approximately 30 percent of total U.S. GHG emissions and is responsible for significant percentages of other U.S. air pollution emissions such as particulate matter, ozone, carbon monoxide, nitrogen oxides, sulfur dioxide, and hazardous air pollutants.[70] Congress, the EPA, the U.S. Department of Transportation (DOT), and the states have enacted numerous initiatives and regulations to reduce these harmful air emissions from the transportation sector. These efforts include increasing vehicle efficiency, mandating the use of transportation fuels with reduced air emissions, and, to a lesser extent, land use planning to reduce vehicle use and encourage alternative forms of transportation such as mass transit, bicycle transit, and walking. In late 2015, five northeastern states and Washington, D.C. announced that they will put in place a new regional market-based cap and trade program to reduce GHG emissions from the transportation sector. These states plan to invest significant resources in new transportation policies, reducing traffic

---

[68] Ladra v. New Dominion, LLC, 353 P.3d 529, 531 (Okla. 2015).

[69] FPL Farming Ltd. v. Envtl. Processing Sys., 351 S.W.3d 306, 310 (Tex. 2011).

[70] *See Sources of Greenhouse Gas Emissions, Transportation Sector Emissions*, U.S. ENVTL. PROT. AGENCY, http://www3.epa.gov/climatechange/ghgemissions/sources/transportation.html; *Cars, Trucks, and Air Pollution*, UNION OF CONCERNED SCIENTISTS, http://www.ucsusa.org/clean-vehicles/vehicles-air-pollution-and-human-health/cars-trucks-air-pollution#VrVO3Sk47UU.

congestion, and increasing clean vehicle use, among other initiatives.[71] Many other states, counties, and municipalities, also have transportation plans in place to reduce emissions. The remainder of this section discusses two of the major regulatory regimes to reduce adverse environmental impacts from the transportation sector—regulation of vehicle emissions and regulation of biofuels.

A. Regulation of Vehicle Emissions

Both the EPA and the DOT regulate air emissions from automobiles and other motor vehicles. Until 2010, these two agencies issued separate sets of regulations under different sources of statutory authority. The CAA regulates auto emissions by requiring the EPA to regulate the emissions of air pollutants from new motor vehicles that may cause or contribute to air pollution that may reasonably be anticipated to endanger public health and welfare.[72] After the U.S. Supreme Court's decision in *Massachusetts v. EPA* (discussed later in this Chapter), the EPA included GHG emissions within the scope of air pollutants to regulate from new motor vehicles. The EPA has enacted different standards to limit GHG emissions from different classes of vehicles, such as passenger vehicles, light-duty trucks, and heavy-duty trucks. Under CAA § 209, Congress has preempted states from setting motor vehicle emissions standards that are more stringent than those set by the federal government, except for California, which may petition EPA for a preemption waiver to adopt its own, stricter standards. Once California adopts a stricter standard, other states may follow it, resulting in the potential for two, rather than fifty, different sets of vehicle emission standards throughout the country.

In addition to the EPA's regulation of auto emissions under the CAA, the National Highway Traffic Safety Administration (NHTSA) within the DOT regulates fleet-wide automobile fuel efficiency standards, known as the Corporate Average Fuel Economy (CAFE) standards.[73] Congress established the CAFE standards in the Energy Policy and Conservation Act of 1975 (EPCA) in response to the Middle East oil crisis in the early 1970s. The original goal of the

---

[71] Amanda Reilly, *5 States, D.C. Launch Market-Based Emissions Initiative*, GREENWIRE, Nov. 24, 2015; GEORGETOWN CLIMATE CENTER, REDUCING GREENHOUSE GAS EMISSIONS FROM TRANSPORTATION: OPPORTUNITIES IN THE NORTHEAST AND MID-ATLANTIC (Nov. 2015).

[72] *See Transportation and Climate, Regulations & Standards*, U.S. ENVTL. PROT. AGENCY, http://www3.epa.gov/otaq/climate/regulations.htm.

[73] *See CAFE—Fuel Economy*, NAT'L HIGHWAY TRAFFIC SAFETY ADMIN., http://www.nhtsa.gov/fuel-economy.

CAFE standards was to double new car fuel efficiency standards from 13.6 miles per gallon (mpg) in 1974 to 27.5 mpg by 1985.

Since 2010, the EPA and NHTSA have engaged in several joint rulemaking proceedings imposing significantly higher CAFE standards as well as limits on GHG emissions. In May 2010 these agencies issued a final rule aimed at reducing GHG emissions from light-duty vehicles.[74] Through the rule, the EPA set "national CO_2 emission standards," requiring light-duty vehicles "to meet an estimated combined average emissions level of 250 grams/mile of CO_2 in model year 2016."[75] Also under this rule, NHTSA set CAFE standards "for passenger cars and light trucks," requiring "manufacturers of these vehicles to meet an estimated combined average fuel economy level of 34.1 mpg in model year 2016."[76] If vehicle manufacturers were to meet both of these requirements just by improving fuel economy, the average mpg achieved for the light-duty vehicle fleet would be 35.5 mpg.[77] Both the EPA and NHTSA standards applied to earlier model years (beginning in model year 2012) but became more stringent over time, with the most stringent standards applying to model year 2016 vehicles.[78] In 2012 these agencies then issued "final rules to further reduce greenhouse gas emissions and improve fuel economy for light-duty vehicles for model years 2017 and beyond."[79] Combined, the 2012 NHTSA and EPA rules would require the light duty vehicle fleet to meet an average 54.5 mpg mark if manufacturers only improved fuel economy in order to meet the standards.[80] NHTSA and the EPA have also enacted heightened CAFE standards and GHG emission standards for heavy-duty trucks.[81]

Auto manufacturers that fail to meet the CAFE standards must pay a civil penalty of "$5.50 per each tenth of a mpg that a manufacturer's average fuel economy falls short of the standard for a given model year multiplied by the total volume of those vehicles in the affected fleet."[82] As NHTSA reports:

---

[74] Light-Duty Vehicle Greenhouse Gas Emission Standards and Corporate Average Fuel Economy Standards, 75 Fed. Reg. 25,324, 25,329 (May 7, 2010).

[75] *Id.* at 25,329–30.

[76] *Id.* at 25,330.

[77] *Id.*

[78] *Id.*

[79] 2017 and Later Model Year Light-Duty Vehicle Greenhouse Gas Emissions and Corporate Average Fuel Economy Standards, 77 Fed. Reg. 62,624, 62,626 (Oct. 15, 2012).

[80] *Id.* at 62,627 n.3.

[81] *See Transportation and Climate, Regulations & Standards: Heavy-Duty*, U.S. ENVTL. PROT. AGENCY, http://www3.epa.gov/otaq/climate/regs-heavy-duty.htm.

[82] 77 Fed. Reg. at 63,126.

Since 1983, manufacturers have paid more than $500 million in civil penalties. Most European manufacturers regularly pay CAFE civil penalties ranging from less than $1 million to more than $20 million annually. Asian and domestic manufacturers have never paid a civil penalty.[83]

The most recent, high profile violation (as well as fraudulent activity) associated with the vehicle efficiency and emissions standards is the disclosure in 2015 that Volkswagen had embedded "defeat devices" in many of their diesel vehicles sold in the United States and elsewhere.[84] The devices sensed when the vehicles were being subject to EPA testing and only controlled for pollutants during the tests. At all other times, the device turned off the pollution control equipment in the vehicle to provide improved vehicle performance and increased emissions of pollutants.

B. Regulation and Use of Biofuels

For decades, the United States has supported the development of alternative liquid fuels, primarily biofuels, as a partial substitute for gasoline in the transportation sector with the goals of achieving energy security, lowering transportation fuel prices, and reducing the harmful air emissions associated with the combustion of gasoline in vehicle engines. The feedstocks for biofuels include corn, grain, grasses, forest residues, crop residues, waste biomass, soy, sugarcane, soybean oil, vegetable oil, and recycled grease. Biofuels consist primarily of (1) ethanol, which is an alcohol fuel made from the sugars found in grains like corn, sorghum, and barley as well as in sugarcane and is blended with gasoline to use in vehicle engines; and (2) biodiesel, which is a "fuel made from vegetable oils, fats, or greases" that can be blended with petroleum-based diesel for use in diesel engines.[85] Certain types of biofuels are seen as "cleaner" than gasoline because they emit fewer GHG emissions and other harmful air pollutants when burned in vehicle engines than petroleum-based gasoline or diesel fuel. As explained below, however, there are environmental costs, including air pollution costs, associated with biofuels, which have driven significant changes in the federal and state laws mandating or incentivizing biofuels.

Although federal biofuel policies and mandates are discussed in detail in Chapter 5, we provide a short summary here before focusing

---

[83] *CAFE Overview—Frequently Asked Questions*, NAT'L HIGHWAY TRAFFIC SAFETY ADMIN., http://lobby.la.psu.edu/_107th/126_CAFE_Standards_2/Agency_Activities/NHTSA/NHTSA_Cafe_Overview_FAQ.htm.

[84] *See Volkswagen Light Duty Diesel Vehicle Violations for Model Years 2009–2016*, U.S. ENVTL. PROT. AGENCY, http://www.epa.gov/vw.

[85] *See Biofuels: Ethanol and Biodiesel Explained*, U.S. ENVTL. PROT. AGENCY, https://www.eia.gov/energyexplained/index.cfm?page=biofuel_home.

on the GHG and other environmental components of these policies. As summarized by the Congressional Research Service, in 2005, Congress enacted the federal Renewable Fuel Standard (RFS) in the Energy Policy Act of 2005.[86] The RFS, which the EPA administers under the CAA, "mandated that a minimum of 4 billion gallons of renewable fuel" be used in the nation's gasoline supply in 2006, and that this amount would increase to 7.5 billion gallons by 2012.[87] Soon after, in 2007, Congress enacted the Energy Independence and Security Act (EISA), which increased those amounts significantly to require the use of "9 billion gallons of biofuels in 2008," increasing to 36 billion gallons in 2022, with a cap of 15 billion gallons from corn ethanol.[88] EISA expanded the RFS to apply to most U.S. transportation fuels—not only gasoline but also diesel fuels used in automotive, non-road, locomotive, and marine engines.[89] EISA also required an increasing amount of the renewable fuel mandate to be met with "advanced" biofuels, which include biomass-based diesel fuel, cellulosic ethanol,[90] and other advanced biofuels produced from a variety of feedstocks other than corn starch. These biofuels emit fewer "lifecycle" GHG emissions than either petroleum or corn ethanol.[91]

To count toward the required amount of advanced biofuels and biomass-based diesel in EISA, the biofuel must have GHG reductions of 50 percent as compared to a 2005 baseline average of GHG emissions from the gasoline or diesel fuel it replaces.[92] Cellulosic ethanol must contain a 60 percent reduction from the baseline to count toward that biofuel's mandated gallon amount.[93] These mandates create a guaranteed market for large amounts of biofuels, acting as an "indirect subsidy for capital investment in the

---

[86] RANDY SCHNEPF & BRENT YACOBUCCI, CONG. RESEARCH SERV., RENEWABLE FUEL STANDARD (RFS): OVERVIEW AND ISSUES 1 (Mar. 14, 2013), https://www.fas.org/sgp/crs/misc/R40155.pdf; Energy Policy Act of 2005, Pub. L. no. 109–58.

[87] SCHNEPF & YACOBUCCI, *supra* note 86, at 1.

[88] *Id.* at 1 & Fig. 1; Energy Independence & Security Act of 2007, Pub. L. no. 110–140.

[89] SCHNEPF & YACOBUCCI, *supra* note 86, at 1–2.

[90] *See* KELSI BRACMORT, CONG. RESEARCH SERV., THE RENEWABLE FUEL STANDARD (RFS): CELLULOSIC BIOFUELS (Jan. 14, 2015) (discussing policies governing cellulosic ethanol, which are biofuels made from grasses, trees, or the inedible or waste parts of plants); *Ethanol Feedstocks*, U.S. DEP'T OF ENERGY, ALTERNATIVE FUELS DATA CTR., http://www.afdc.energy.gov/fuels/ethanol_feedstocks.html.

[91] BRACMORT, *supra* note 90, at 1; *see also* 42 U.S.C. § 7545(o)(1)(H) (West 2016) (discussing lifecycle GHG emissions); Regulation of Fuels and Fuel Additives: Changes to Renewable Fuel Standard Program, 75 Fed. Reg. 14,670, 14,670 (Mar. 26, 2010) ("This rulemaking [implementing EISA] marks the first time that greenhouse gas emission performance is being applied in a regulatory context for a nationwide program."). For further discussion of lifecycle emissions, see *infra* notes 98 and 109 and accompanying text.

[92] 75 Fed. Reg. 14,670, *supra* note 91, at 14,677.

[93] *Id.*

construction of biofuels plants" and for the industry as a whole.[94] Corn ethanol qualifies toward the total renewable fuel mandate if it reduces lifecycle GHG emissions by at least 20 percent relative to the 2005 baseline average GHG emissions from the gasoline or diesel fuel it replaces.[95] But if a renewable fuel facility commenced construction prior to December 2007 (the effective date of EISA), or an ethanol facility was constructed prior to December 31, 2009 and is powered with natural gas or biomass, fuel produced at those plants is "grandfathered" and counts toward the total renewable fuel amounts required by EISA regardless of GHG emissions.[96]

EPA has issued comprehensive rules to implement the RFS. It issued the first set—RFS1—after the enactment of the initial 2005 RFS. It enacted the second set—RFS2—in connection with EISA.[97] In addition, in EISA, Congress tasked EPA with creating a lifecycle analysis of GHG emissions for each category of renewable fuel in order to ensure that the lifecycle of each type of renewable fuel results in a specific percentage reduction of GHG emissions as compared to the gasoline or diesel fuel it replaces. A lifecycle analysis for carbon emissions for fuels includes the emissions from the production or consumption of the fuel in vehicles, the emissions associated with transporting the fuel to the source of consumption, the emissions associated with producing the fuel, and the emissions associated with changing the land use to produce the feedstock. Many argue that the RFS so far has not met the goal of encouraging the development of environmentally friendly fuels.[98] To date, almost all of the RFS is met by corn-based ethanol.[99] While corn-based ethanol is "renewable" in the sense that corn is not a finite resource like oil, coal, or natural gas, some studies have shown that the CO_2 and other GHG emissions associated with the life-cycle production of corn-based ethanol are significant and in some cases approach or exceed that of gasoline.[100]

---

[94] SCHNEPF & YACOBUCCI, *supra* note 86, at 2.

[95] 75 Fed. Reg. 14,670, *supra* note 91, at 14,677; SCHNEPF & YACOBUCCI, *supra* note 86, at 4.

[96] 75 Fed. Reg. 14,670, *supra* note 91, at 14,677, 14,688, 14,787–91.

[97] 75 Fed. Reg. 14,670, *supra* note 91, at 14,670; SCHNEPF & YACOBUCCI, *supra* note 86, at 2.

[98] *See generally* Melissa Powers, *King Corn, Will the Renewable Fuel Standard Eventually End Corn Ethanol's Reign?*, 11 VT. J. ENVTL. L. 667, 668–74 (2010).

[99] Roberta F. Mann & Mona L. Hymel, *Moonshine to Motorfuel: Tax Incentives for Fuel Ethanol*, 19 DUKE ENVTL. L. & POL'Y F. 43, 46–47 (2008); *U.S. Ethanol Exports Exceed 800 Million Gallons for Second Year in a Row*, U.S. ENERGY INFO. ADMIN., March 10, 2016, https://www.eia.gov/todayinenergy/detail.cfm?id=25312 ("Corn is the primary feedstock of ethanol in the United States, and large corn harvests have contributed to increased ethanol production in recent years.").

[100] *See* Powers, *supra* note 98, at 669 (describing the studies).

Several states, most notably California, have gone beyond the RFS to rely more heavily on life cycle analysis in the transportation fuels sector to limit GHG emissions of all fuels sold in the state. In 2006, the California Legislature enacted AB 32, the California Global Warming Solutions Act. As part of AB 32, the legislature made extensive findings about the connection between GHG emissions and global warming and, specifically, the threat global warming poses for California, including sea level rise, reduction in water quality and quantity, and harm to public health and the environment.[101] The legislature set a target of reducing the state's GHG emissions to 1990 levels by 2020, with further reductions beyond 2020. It also directed the California Air Resources Board (CARB) to adopt rules and regulations "to achieve the maximum technologically feasible and cost-effective greenhouse gas emissions reductions."[102]

Because transportation emissions are a significant contributor to GHG emissions in California, the state, through an executive order and subsequent regulations, limited the carbon content of transportation fuels. As explained by the federal court that has affirmed the validity of these regulations (with certain issues left open upon remand), the Governor "directed CARB to adopt regulations that would reduce the average GHG emissions attributable to California's fuel market by 10 percent by 2020."[103] In response, CARB created the Low Carbon Fuel Standard (LCFS), which, as summarized in previous work, established a "baseline, average carbon intensity for all vehicular fuels consumed in California" (using the average carbon intensity of the 2010 gasoline market) and required "each supplier of vehicular transportation fuels in the state to reduce its average carbon intensity from that baseline by set amounts each year between 2011 and 2020."[104] The LCFS also allows for "suppliers to generate credits for exceeding the reduction required for that year,"[105] permitting those credits to be used to offset deficits or to sell to other blenders, thus creating a market for trading, banking, and borrowing of credits.[106] CARB's intent was to form a market that would create incentives for the development of

---

[101] 2006 Cal. Legis. Serv. Ch. 488 (A.B. 32).

[102] *Id.*

[103] Rocky Mountain Farmers Union v. Corey, 730 F.3d 1070, 1080 (9th Cir. 2013); CAL. EXEC. ORDER No. S–01–07 (Jan. 18, 2007).

[104] Alexandra B. Klass, *Tax Benefits, Property Rights, and Mandates: Considering the Future of Government Support for Renewable Energy*, 20 J. ENVTL. & SUSTAINABILITY L. 19 (2013); *see also* Rocky Mountain Farmers Union, 730 F.3d at 1080; CAL. CODE REGS. tit. 17, §§ 95480–95482 (West 2016). In 2011, the carbon intensity cap was set 0.25 percent below the 2010 average with further reductions each year between 2011 and 2020. *Rocky Mountain Farmers Union*, 730 F.3d at 1082.

[105] Klass, *supra* note 104, at 66.

[106] *Rocky Mountain Farmers Union*, 730 F.3d at 1080.

low-carbon fuels for sale in the California market and allow the state to meet the requirements of AB 32.[107]

As the court further explains, like the federal RFS, the California LCFS adopts a lifecycle analysis for carbon emissions for fuels that includes the emissions from: (1) conversion of the land to agricultural use; (2) the growth and transportation of the feedstock (corn, sugar, other plant matter) with credit for the GHGs absorbed during photosynthesis; (3) the process used to convert the feedstock into liquid fuel and the efficiency of that process; (4) the source of electricity used to power the production plant (coal, natural gas, wind, nuclear, etc.); (5) the fuel used for thermal energy; and (6) transportation distance of the fuel to the blender in California and the form of transportation (truck, train, ship, etc.).[108]

These factors can both help and harm ethanol from different geographic regions. For example, most corn is grown in the Midwest, and the emissions generated by transporting corn from fields to ethanol plants in the Midwest are relatively low as compared to transporting the corn to more distant plants. But transporting the ethanol from the Midwest to California is more carbon intensive than transporting ethanol from California plants.[109] Likewise, corn ethanol produced in California or sugarcane ethanol produced in Brazil generally relies on lower-carbon electricity than ethanol produced in the Midwest, as California's electric grid runs largely on natural gas and renewable energy, Brazil uses primarily hydropower, and the Midwest relies more heavily on coal.[110]

California's LCFS differs from the RFS in that it addresses not just biofuels but all transportation fuels, including gasoline, diesel, hydrogen, natural gas, and electricity, and requires producers to reduce the lifecycle GHG emissions of all fuels over time.[111] Also, unlike the federal RFS, California's LCFS does not create fuel categories or minimum lifecycle reductions, but instead considers the carbon emissions of each fuel separately, and rewards all reductions in lifecycle emissions.[112]

---

[107] *Id.* at 1080.

[108] *Id.* at 1083.

[109] Alexandra B. Klass & Elizabeth Henley, *Energy Policy, Extraterritoriality, and the Dormant Commerce Clause*, 5 SAN DIEGO J. CLIMATE & ENERGY L. 127, 160 (2013–2014).

[110] *Rocky Mountain Farmers Union*, 730 F.3d at 1083.

[111] PROMOTUM, CALIFORNIA'S LOW CARBON FUEL STANDARD: EVALUATION OF THE POTENTIAL TO MEET AND EXCEED THE STANDARDS (2015), http://www.ucsusa.org/sites/default/files/attach/2015/02/California-LCFS-Study.pdf.

[112] For more detail about the California LCFS regulations and implementation, see Klass & Henley, *supra* note 109.

Perhaps predictably, fuel producers—including ethanol producers and oil companies—challenged the LCFS, suing California in 2009 and arguing that the LCFC violated the dormant Commerce Clause of the U.S. Constitution. A dormant Commerce Clause violation can occur in three different ways. First, a state statute or regulation can unlawfully regulate extraterritorial conduct, meaning that it imposes mandates wholly outside of the boundaries of the state. Second, the state law can discriminate against interstate commerce on its face or in practice or effect, meaning that the actual text of the law makes clear that the statute favors in-state commerce over out-of-state commerce or the statute causes this type of favoritism. Finally, the state law can have indirect effects on out-of-state commerce, and if these effects on interstate commerce are sufficiently strong as compared to the localized benefits produced by the statute, this can also violate the dormant Commerce Clause.

The plaintiffs in the LCFS case raised all three of these issues. The Ninth Circuit ultimately held that the LCFS did not regulate extraterritorially, concluding as follows:

> [The LCFS] says nothing at all about ethanol produced, sold, and used outside California, it does not require other jurisdictions to adopt reciprocal standards before their ethanol can be sold in California, it makes no effort to ensure the price of ethanol is lower in California than in other states, and it imposes no civil or criminal penalties on non-compliant transactions completed wholly out of state.[113]

The court also found that the LCFS did not discriminate against interstate commerce on its face because, although some of the factors used to calculate the carbon intensity of fuels appeared to harm out-of-state sources—for example, the factor that addressed emissions from transporting fuel to California—the LCFS considered of a range of factors, all of which were used to calculate carbon intensity of fuels. The court concluded that "[e]ach factor in the default pathways is an average based on scientific data, not an ungrounded presumption that unfairly prejudices out-of-state ethanol."[114] Further, the court noted that a state regulation "is not facially discriminatory simply because it affects in-state and out-of-state interests unequally."[115] The regulation simply must have a reason—based on something other than the origin of the products—for treating products

---

[113] *Rocky Mountain Farmers Union*, 730 F.3d at 1102–03.
[114] *Id.* at 1089.
[115] *Id.*

differently, and the LCFS treats fuels differently based on their carbon intensity, not their origin.[116]

Additionally, the court found that for oil, the LCFS did not discriminate against interstate commerce in purpose or effect. Plaintiffs had argued that the rule, as applied, favored a specific type of California oil for purposes of economic protectionism, but the court disagreed.[117] The court remanded the case to the federal district court for two additional considerations: whether the LCFS discriminates against out-of-state ethanol (as opposed to oil) in purpose or effect, and whether it has an impermissible indirect effect on out-of-state ethanol interests.[118] The federal district court is still in the process of considering these issues as well as additional dormant Commerce Clause issues that arose after California made certain changes to the original LCFS—issues that the Ninth Circuit partially addressed on the merits and partially remanded.[119]

Since California's enactment of the LCFS, Governor Jerry Brown has issued an executive order to reduce petroleum use in the state by 40 percent by 2030, and the LCFS will be an integral component to reaching that goal.[120] Moreover, Oregon and Washington have begun to put into place similar laws to reduce the GHG emissions of transportation fuels.

V. Environmental Regulation of Air Pollution from the Electric Power Sector[121]

A. Federal Environmental Laws and Regulations Governing the Electric Power Sector

The electric power sector produces a significant amount of air pollution, primarily because it relies so heavily on coal-fired power, which is one of the largest sources of air pollution in the United States. The combustion of coal releases a range of air pollutants including sulfur, nitrogen, particulate matter, and hazardous air pollutants such as mercury.[122] The electric power sector also

---

[116] *Id.*

[117] *Id.* at 1098–99.

[118] *Id.* at 1107.

[119] *See* American Fuels & Petrochemical Ass'n v. Corey, 2015 WL 5096279 at 38–39 (E.D. Cal., Aug. 28, 2015) (dismissing some, but not all, of plaintiffs' claims).

[120] *Governor Brown Establishes Most Ambitious Greenhouse Gas Reduction Target in North America*, OFFICE OF GOVERNOR EDMUND G. BROWN, JR., Apr. 29, 2015, https://www.gov.ca.gov/news.php?id=18938.

[121] A wide range of federal, state, and local environmental laws also govern water, hazardous waste, and hazardous substance discharges from electric power plants. These laws are beyond the scope of this Chapter.

[122] *See* JAMES E. MCCARTHY, CONG. RESEARCH SERV., CLEAN AIR ISSUES IN THE 113TH CONGRESS: AN OVERVIEW 3 (Jan. 4, 2013).

represents over 30 percent of U.S. GHG emissions.[123] The Clean Air Acts of 1963, 1970, and amendments of 1977 and 1990, form the backbone of U.S. air pollution control efforts, with a significant emphasis on regulating emissions from coal-fired power plants.

Power plants are one of the main emitters of several air pollutants of particular concern. Power plants cause approximately 73 percent of all sulfur dioxide emissions in the United States,[124] although their contribution to SO_2 is declining as the use of coal-fired power declines.[125] The EPA notes that SO_2 has adverse respiratory effects on humans and that children, the elderly, and persons suffering from asthma are particularly sensitive to SO_2.[126] SO_2 also causes acid rain, which harms forests, aquatic life, and other resources, and it damages crops and buildings. Power plants are also large emitters of nitrogen oxides (NO_x),[127] and, like SO_2, NO_x causes lung problems in humans and contributes to acid rain.[128] Further, power plants emit approximately 50 percent of all mercury emissions in the United States.[129] This pollutant can have "toxic effects on the nervous, digestive and immune systems, and on lungs, kidneys, skin and eyes"; it also builds up in the systems of fish and harms birds and other wildlife that eat the fish.[130] And finally, power plants are the largest single source category of these emissions, surpassing industry, transportation, agriculture, and commercial and residential sources.[131]

The EPA, operating under its existing CAA authority, has taken on the bulk of the responsibility for addressing these emissions, although Congress—at least historically—provided assistance. To address acid rain problems, in particular, through the CAA Amendments of 1990 Congress added a new cap-and-trade program to the Act, creating an overall cap on SO_2 emissions and a system in

---

[123] *See Sources of Greenhouse Gas Emissions, Electricity Sector Emissions*, U.S. ENVTL. PROT. AGENCY, http://www3.epa.gov/climatechange/ghgemissions/sources/electricity.html.

[124] *Sulfur Dioxide Basics*, U.S. ENVTL. PROT. AGENCY, https://www.epa.gov/so2-pollution/sulfur-dioxide-basics#what%20is%20so2.

[125] *Power Plant Emissions of Sulfur Dioxide and Nitrogen Oxides Continue to Decline in 2012*, U.S. ENERGY INFO. ADMIN., Feb. 27, 2013, http://www.eia.gov/todayinenergy/detail.cfm?id=10151.

[126] *Sulfur Dioxide, Basics*, U.S. ENVTL. PROT. AGENCY, *supra* note 124.

[127] *Nitrogen Oxides*, U.S. ENVTL. PROT. AGENCY, https://www3.epa.gov/cgi-bin/broker?polchoice=NOX&_debug=0&_service=data&_program=dataprog.national_1.sas.

[128] U.S. ENVTL. PROT. AGENCY, NO_x: HOW NITROGEN OXIDES AFFECT THE WAY WE LIVE AND BREATHE at 3 (1998).

[129] *Mercury and Air Toxic Standards*, U.S. ENVTL. PROT. AGENCY, https://www.epa.gov/mats.

[130] *Mercury and Health*, WORLD HEALTH ORG., http://www.who.int/mediacentre/factsheets/fs361/en/.

[131] *Id.*

which each power plant would be allotted a certain amount of allowances, each representing a unit of emissions. Power plants that could easily reduce their emissions below the allowances they had received sold their extra allowances to plants that had trouble meeting their limit, thus creating compliance flexibility. This portion of the CAA also placed limits on NO_x emissions. Since 1990 Congress has done little else to address power plant emissions, thus leaving the EPA to pick up the bulk of the work.

The EPA further addressed SO_2 and NO_x emissions through a rule called the "NO_x SIP Call" issued in 1998.[132] Under the CAA, the EPA sets total acceptable levels of certain types of pollutants in the air, and states are required to write "State Implementation Plans" (SIPs) to ensure that pollutants within states remain at or below these levels. If state SIPs fail to achieve these standards, the EPA may essentially "recall" the SIPs and impose a Federal Implementation Plan. Through the NO_x SIP Call, the EPA attempted to address the growing problem of states allowing NO_x sources within their state—especially power plants—to emit NO_x that drifted downwind and affected other states. Under the "good neighbor" provision of the Clean Air Act, states within their SIPs are not allowed to permit sources within their state to "contribute significantly" to other states' air quality problems or interfere with other states' achievement of air quality standards.[133] The NO_x SIP Call required states to change their SIPs to address these emissions and attempted to aid the states in reducing NO_x emissions by establishing NO_x "budgets" for each state, which calculated the amount of NO_x emissions that would be allowed within each state after the state required sources to reduce NO_x to the extent that sources could achieve these reductions in a "highly cost-effective" manner. Thus, each state's NO_x budget represented, essentially, a state-level cap on acceptable NO_x emissions from sources within the state—a cap that took into account the current emissions within the state and the amount by which these emissions could be reduced through cost-effective controls.[134] Under the rule, if states elected to allow trading and implement a NO_x trading program, sources in states that had difficulty meeting their cap could trade with sources

---

[132] Finding of Significant Contribution and Rulemaking for Certain States in the Ozone Transport Assessment Group Region for Purposes of Reducing Regional Transport of Ozone, 63 Fed. Reg. 57,356 (Oct. 27, 1998).

[133] 42 U.S.C. § 7410(a)(2) (West 2016). *See also* Michigan v. EPA, 213 F.3d 663, 669 (D.C. Cir. 2000) (describing the rule and the good neighbor provision).

[134] *Michigan v. EPA*, 213 F.3d at 681–82 (describing the budgets).

in states that would over-comply.[135] Legal challenges to the rule failed, although three states were removed from the rule.[136]

The NO_x SIP Call failed to fully address the problem of NO_x emissions flowing interstate and causing air quality problems, and the EPA in 2005—using the same "good neighbor" provision of the CAA—attempted to replace it, issuing another rule that aimed to further curtail NO_x emissions that contributed to downwind problems. The Clean Air Interstate Rule (CAIR) was similar to the NO_x SIP Call in that it required state SIP revisions to reduce SO_2 and NO_x, created state budgets for these emissions, and allowed interstate trading.[137] However, the D.C. Circuit vacated the rule and remanded to the EPA, finding that the rule—particularly because it allowed trading—failed to ensure that within each state sources would avoid contributing significantly to downwind pollution or interfering with downwind states' maintenance of pollution standards.[138] In the meantime, the NO_x SIP Call, which was similar to CAIR, remained in place.[139]

The EPA made a third effort to address interstate SO_2 and NO_x air quality problems through the Cross-State Air Pollution Rule, issued in 2011.[140] This rule was similar to the NO_x SIP Call and CAIR—also setting individual state budgets for NO_x and SO_2—but the EPA more carefully calculated each state's contribution to downwind pollution[141] and created a trading program that prevented trading that would cause a state to violate its mandate to avoid significant contributions to downwind problem.[142] The U.S. Supreme Court affirmed the validity of this rule in 2014.[143]

The EPA also had to make more than one attempt to address mercury pollution from power plants through rules issued under the CAA. Mercury is regulated as a special type of pollutant—a hazardous air pollutant (HAP)—under the CAA, and the EPA must

---

[135] *Id.* at 686 (summarizing the trading program).

[136] *Id.* at 695.

[137] Rule to Reduce Interstate Transport of Fine Particulate Matter and Ozone (Clean Air Interstate Rule); Revisions to Acid Rain Program; Revisions to the NOx SIP Call, 70 Fed. Reg. 25,162 (May 12, 2005); *see also* North Carolina v. EPA, 531 F.3d 896, 903 (D.C. Cir. 2008) (summarizing the rule).

[138] *North Carolina v. EPA*, 531 F.3d at 908, 930.

[139] *Id.*

[140] Federal Implementation Plans: Interstate Transport of Fine Particulate Matter and Ozone and Correction of SIP Approvals, 76 Fed. Reg. 48,208 (Aug. 8, 2011).

[141] EPA v. EME Homer City Generation, 134 S. Ct. 1584, 1596–98 (2014) (summarizing the rule).

[142] *Cross-State Air Pollution Rule*, U.S. ENVTL. PROTECTION AGENCY, https://www3.epa.gov/crossstaterule/basic.html (describing the "quality-assured trading program").

[143] *EME Homer City Generation*, 134 S. Ct. at 1610.

follow certain congressional requirements for regulating HAPs when it issues rules for mercury. The CAA lists certain pollutants emitted by certain sources as HAPs,[144] requires the EPA to periodically consider adding pollutants to this list,[145] and requires the EPA to regulate these pollutants. Mercury is a listed pollutant, and when the EPA endeavored to reduce mercury emissions from power plants in 2005, it determined that the best approach would be to implement a type of cap and trade program that would not fit with the regulatory approach for HAPs required by the CAA. The EPA therefore "delisted" mercury emitted from power plants from the CAA and proceeded to issue a cap-and-trade rule for mercury called the Clean Air Mercury Rule (CAMR).[146] The D.C. Circuit vacated the delisting of mercury from power plants that was a prerequisite to CAMR, determining that the EPA had failed to make the necessary findings to support a delisting.[147]

The EPA then proceeded to regulate mercury emissions within the confines of the CAA, which dictates that for sources of HAPs like mercury, the EPA must require the sources to limit emissions to the lowest possible emissions amount that can be accomplished through the application of the "maximum achievable control technology" (MACT) to the pollution source. Using MACT, the EPA set emission limits for mercury emitted from coal-and oil-fired power plants in 2012 through a rule called the "Mercury and Air Toxics" (MATS) rule.[148] The agency estimated that the final rule would "yield annual monetized benefits (in 2007 dollars) of between $37 to $90 billion using a 3 percent discount rate and $33 to $81 billion using a 7 percent discount rate" and that the benefits "outweigh costs by between 3 to 1 or 9 to 1."[149] However, the EPA—although it thoroughly estimated the costs and benefits of the rule—did not rely directly on these cost and benefits in setting mercury emissions limits, believing that it was not supposed to take costs into account in setting a health-based air quality standard for mercury.[150] The

---

[144] 42 U.S.C. § 7412(b) (West 2016).

[145] 42 U.S.C. § 7412(b)(2), (d) (West 2016).

[146] Standards of Performance for New and Existing Stationary Sources: Electric Utility Steam Generating Units, 70 Fed. Reg. 28,606 (May 18, 2005); *see also* New Jersey v. EPA, 517 F.3d 574, 577–78 (D.C. Cir. 2008) (describing the delisting and the rule).

[147] New Jersey v. EPA, 517 F.3d at 584.

[148] National Emission Standards for Hazardous Air Pollutants from Coal- and Oil-Fired Electric Utility Steam Generation Units and Standards of Performance for Fossil-Fuel-Fired Electric Utility, Industrial-Commercial-Institutional, and Small Industrial-Commercial Institutional Steam Generating Units, 77 Fed. Reg. 9304, 9306 (Feb. 16, 2012).

[149] *Id.* at 9305–06.

[150] Michigan v. EPA, 135 S. Ct. 2699, 2709–10 (2015).

U.S. Supreme Court held that the EPA unreasonably interpreted the CAA when it "deemed cost irrelevant to the decision to regulate power plants," and it remanded the case to the D.C. Circuit.[151] On remand from the Supreme Court, the D.C. Circuit remanded the case to the EPA but did not vacate the MATS rule,[152] thus leaving it in place while the EPA prepares an updated finding to support the MATS rule that was more explicitly based on costs.

The final major pollutant emitted by power plants—GHG emissions—has also spurred numerous EPA regulations and court challenges to these regulations. EPA regulation of GHG emissions began after the 2005 case *Massachusetts v. EPA*, in which the U.S. Supreme Court determined that GHG emissions were an "air pollutant" under the CAA.[153] With respect to air pollutants, the CAA provides (in the context of pollutants from motor vehicles, which were at issue in the case):

> [T]he [EPA] Administrator shall by regulation prescribe (and from time to time revise) in accordance with the provisions of this section, standards applicable to the emission of any air pollutant from any class or classes of new motor vehicles or new motor vehicle engines, which in his judgment cause, or contribute to, air pollution which may reasonably be anticipated to endanger public health or welfare.[154]

The Court found that the EPA had "offered no reasoned explanation for its refusal to decide whether greenhouse gases cause or contribute to climate change" and thus had to decide whether GHGs—a particular type of air pollutant—had to be regulated under the CAA.

Following the *Massachusetts v. EPA* decision EPA completed an "endangerment finding" and a "cause or contribute" finding for GHGs emitted from motor vehicles in which it determined that GHGs from motor vehicles cause or contribute to air pollution that could endanger public health or welfare.[155] The EPA then proceeded to regulate CO_2 emissions from vehicles, as described above, and it has since regulated CO_2 emissions from new and modified or reconstructed power plants (those that commence construction on or

---

[151] *Id.* at 2712.

[152] White Stallion Energy Ctr. v. EPA, Case No. 12–1100 (D.C. Cir., Dec. 15, 2015).

[153] Massachusetts v. EPA, 549 U.S. 497, 528–29 (2007).

[154] *Id.* at 506 (quoting 42 U.S.C. § 7521(a)(1)).

[155] *Endangerment and Cause or Contribute Findings for Greenhouse Gases under Section 202(a) of the Clean Air Act*, U.S. ENVTL. PROT. AGENCY, https://www3.epa.gov/climatechange/endangerment/.

after June 18, 2014[156]) and from existing power plants—those for which construction already had begun as of January 8, 2014. The EPA's carbon rule for new, modified, and reconstructed power plants requires newly-constructed coal plants to limit their emissions of CO_2 to 1,400 pounds of CO_2 per gross megawatt-hour of electricity generated (MWH-g).[157] New, modified, and reconstructed natural gas-fired plants that are used for baseload (constant) electricity generation may emit up to 1,000 pounds of CO_2 per MWh-g, and natural gas plants not used for baseload generation may only emit 120 pounds of CO_2 per million British Thermal Units of energy produced.[158]

The EPA's regulation of carbon emissions from existing power plants, which the agency issued on the same day as the rule for new sources, is more complex. This regulation, referred to as the "Clean Power Plan" (CPP), also establishes emission guidelines for individual sources—similar to the rule for new sources, although states may choose to meet the CPP in ways other than imposing these limits on individual sources. The limits are 1,305 pounds of CO_2 per MWh of electricity generated by existing coal-fired power plants, and 771 pounds of CO_2 per MWh of electricity generated by natural gas-fired plants.[159] The EPA set these limits based on estimates of the best emission reductions that each of these types of power plants could achieve. It closely investigated utilities and other generators that operate power plants within the three closely-interconnected portions of the U.S. grid—the Texas, Eastern, and Western Interconnections—and looked at how these entities could achieve carbon reduction by relying more on their lower carbon-plants and less on their higher-carbon plants, either by running their lower-carbon plants more or offsetting the need to run their higher-carbon plants by, for example, building new renewable plants. (Utilities often own multiple generating plants and "switch" generation among these plants for many reasons, relying more or less on certain plants in their inventory to achieve environmental goals, reliability requirements, and other goals and mandates.) The EPA then took the *lowest* amount of emissions reductions that could be achieved by gas-fired plants and coal-fired plants in each of the three interconnections and set the emissions guideline for each type of plant accordingly.

---

[156] Standards of Performance for Greenhouse Gas Emissions from New, Modified, and Reconstructed Stationary Sources: Electric Utility Generating Units, 80 Fed. Reg. 64,510, 64,512 (Oct. 23, 2015).

[157] *Id.* at 64,512.

[158] *Id.* 64,512–13.

[159] Carbon Pollution Emission Guidelines for Existing Stationary Sources: Electric Utility Generating Units, 80 Fed. Reg. 64,662, 64,667 (Oct. 23, 2015).

Based on these individualized emission guidelines, the EPA then set carbon limit "goals" for each state. The EPA identified the number of coal-fired and natural gas-fired plants in each state, estimated how many MWh of electricity these plants generate in a given year, and multiplied the number of plants by the MWh, and then by the emission guidelines, to estimate a total emission limit for each state. States are then required to write plans to accomplish their state goal. Within these plans, states can choose to require that each individual plant meet the emission guideline established in the CPP. Alternatively, they can choose to use a variety of strategies to meet the state carbon goal, including allowing extensive trading among sources within the state and in different states. The EPA anticipates that states might apply three specific "building blocks" to ensure that the states meet their CPP goals, including having utilities and other generators: (1) improve the efficiency of existing coal-fired power plants, (2) rely less on coal-fired power plants and more on existing natural gas-fired plants for the electricity that they generate, and (3) rely more on newly-built renewable energy plants and less on existing coal-fired power plants.[160] States also may rely on nuclear power and energy efficiency, among other tools, to help meet their goals, although they must demonstrate that reliance on these tools is causing reductions in emissions from the regulated sources—coal-and natural gas-fired power plants—by offsetting the need for electricity generation from those sources and thus reducing generation (and emissions) from those sources. States must meet interim carbon reduction goals by 2022, and by 2030 they must have fully achieved their CPP goal.[161] The CPP does not apply to Alaska, Guam, Hawaii, Puerto Rico, Washington, D.C., and Vermont, because Alaska, Guam, Hawaii, and Puerto Rico do not have electric grids connected to the national grid, and Washington, D.C. and Vermont do not have any CPP-regulated power plants.[162]

Industry groups and numerous states challenged the CPP, and in February 2016 the U.S. Supreme Court stayed the enforcement of the rule while the cases challenging the CPP before the D.C. Circuit Court of Appeals are pending.[163] Although approximately 27 states have challenged the CPP, approximately 18 states as well as the "District of Columbia, five cities, and a county" have intervened in

---

[160] *See* Hannah J. Wiseman & Hari M. Osofsky, *Regional Energy Governance and U.S. Carbon Emissions*, 43 ECOLOGY L.Q. 143, 159 (2016) (summarizing the building blocks).

[161] *Id.*

[162] 80 Fed. Reg. 64,662, *supra* note 159, at 64,708.

[163] West Virginia v. EPA, 136 S. Ct. 1000 (2016).

support of the plan.[164] Some states are continuing to write their plans for complying with the CPP despite the stay.[165]

B. State Environmental Laws Governing the Electric Power Sector

In addition to the federal environmental laws governing air emissions from the electric power sector, a variety of state laws impose complementary and in many cases more stringent requirements on these air emissions. As a result of the cooperative federalism approach embodied in the Clean Air Act, states work closely with the EPA to reduce a wide range of criteria air pollutants and toxic air pollutants from the electric power sector. And even though Congress has not enacted specific legislation to address GHG emissions from the electric power sector to date, numerous states have done so. These state laws are discussed below.

CO_2 Emission Performance Standards—As explained by the EPA:

> As of March 2014, four states—California, New York, Oregon and Washington—have enacted mandatory GHG emission standards that impose enforceable emission limits on new and/or expanded electric generating units. Three states—California, Oregon and Washington—have enacted mandatory GHG emission performance standards that set an emission rate for electricity purchased by electric utilities.[166]

For the GHG emission standards, generating units must "measure and report on electricity generation and CO_2 emissions on a regular basis to verify their compliance with the standards."[167]

Energy Efficiency Policies—The EPA further describes programs implemented by states to reduce consumers' demand for energy and thus to lower the need for electricity generation, which, in turn, reduces pollutant emissions from this generation. The agency notes:

> States have employed a variety of strategies to increase investment in demand-side energy efficiency technologies and practices, including (1) energy efficiency resource standards [which require utilities to implement programs that achieve a certain amount of energy efficiency], (2) demand-side energy efficiency programs, (3) building energy codes, (4) appliance standards [when not preempted

---

[164] Wiseman & Osofsky, *supra* note 160, at 208.

[165] *Id.* at 209.

[166] U.S. ENVTL. PROT. AGENCY, SURVEY OF EXISTING STATE POLICIES AND PROGRAMS THAT REDUCE POWER SECTOR CO_2 EMISSIONS 17-18 (June 2, 2014).

[167] *Id.* at 19.

by U.S. Department of Energy Standards] and (5) tax credits.[168]

We further discuss energy efficiency in Chapter 8.

Renewable Portfolio Standards (RPSs)—As discussed in more detail in Chapter 5, over 25 states have enacted RPSs, which require covered electric utilities in the state to procure a certain percentage of electricity sales (10 percent, 15 percent, 30 percent, etc.) from designated renewable energy resources (wind, solar, geothermal, etc.) by a certain date (2015, 2020, 2030, etc.).[169] State RPSs vary in terms of which entities must comply—for example, all utilities in the state or just large utilities; the date by which the percentage target must be achieved; the types of renewable energy that are covered by the standard and can be used by utilities to achieve the standard; and how much renewable energy must come from in-state sources.[170]

California AB 32—As discussed earlier in this Chapter, in 2006, California enacted the Global Warming Solutions Act, also known as California AB 32.[171] As explained by the California agency that implements the Act, "AB 32 requires the state to reduce its GHG emissions to 1990 levels by 2020 (approximately 15%)."[172] AB 32 directed the California Air Resources Board (CARB) to achieve these reductions from the electric power sector, the transportation sector, and other sources. CARB is addressing transportation sector emissions through its LCFS discussed earlier in this Chapter. As for the electricity sector, the state imposed a 33 percent RPS on electric utilities to reduce GHG emissions from that sector by 2020. It also placed a total cap on GHG emissions that declines by 2 percent each year, coupled with an emissions trading system through which emitters of GHG emissions must obtain allowances based on the tons of GHGs it emits. Participants are given allowances to start and must stay below that allowance amount or purchase necessary allowances at a quarterly auction or from an entity with excess allowances. The cap and trade program applies to utilities, large industrial plants, and fuel distributors. California intends to link its program with other western states and Canadian provinces through the Western Climate Initiative, discussed below. Since AB 32 was enacted, Governor Jerry Brown has declared new emissions reduction targets of 40 percent below 1990 levels by 2030 and 80% below 1990 levels

---

[168] *Id.* at 23.

[169] *See* Database of State Incentives for Renewables & Efficiency, Renewable Portfolio Standard Policies (June 2016) (summary map).

[170] HARRISON FELL ET AL., RESOURCES FOR THE FUTURE, DESIGNING RENEWABLE ELECTRICITY POLICIES TO REDUCE EMISSIONS 13 (2012).

[171] *Assembly Bill 32 Overview*, CAL. AIR RES. BD., http://www.arb.ca.gov/cc/ab32/ab32.htm.

[172] *Id.*

by 2050.[173] In 2015, California expanded its RPS to require electric utilities to obtain 50% of retail sales from eligible renewable energy resources by 2050.[174]

State bans on new coal-fired generation—California, Minnesota, Washington, and Oregon have enacted legislation limiting or prohibiting utilities in the state from generating electricity from coal-fired power plants constructed after the effective date of the legislation and, in some cases, prohibiting the import of coal-fired power to the state.[175] The State of North Dakota, electric cooperatives, and the North Dakota lignite coal industry challenged the Minnesota law on dormant Commerce Clause and preemption grounds. In 2016, in a split opinion, the U.S. Court of Appeals for the Eighth Circuit invalidated the portion of the law that banned coal-fired electricity imports, with one judge finding that the law regulated extraterritorially in violation of the dormant Commerce Clause and the two other panel judges invalidating the law solely on federal preemption grounds.[176]

C. Regional Collaborations to Limit GHG Emissions from the Electric Power Sector

The Regional Greenhouse Gas Initiative (RGGI), created in 2005, was the "first mandatory cap-and-trade program in the United States to limit carbon dioxide."[177] As of 2016, Connecticut, Delaware, Maine, Maryland, Massachusetts, New Hampshire, New York, Rhode Island, and Vermont participate in the initiative. As the Center for Climate and Energy Solutions explains, "RGGI sets a cap on CO_2 emissions from power plants throughout the region, and allows regulated entities to trade carbon emission allowances to achieve compliance."[178] The Center estimates that "[b]etween 2009 and 2013, carbon emissions from power plants in the RGGI region decreased by 45 percent as a result of fuel switching to natural gas, increased use of renewable energy, and a reduction in regional energy consumption."[179] The success of the program has allowed the states to establish more ambitious goals: "Following a 2012 design

---

[173] *Governor Brown Establishes Most Ambitious Greenhouse Gas Reduction Target in North America, supra* note 120.

[174] *California RPS*, DATABASE OF STATE INCENTIVES FOR RENEWABLES & EFFICIENCY, http://programs.dsireusa.org/system/program/detail/840.

[175] U.S. ENVTL. PROT. AGENCY, *supra* note 166, at 17.

[176] *See* North Dakota v. Heydinger, 825 F.3d 912 (8th Cir. 2016).

[177] *Regional Greenhouse Gas Initiative: Summary*, CTR. FOR CLIMATE & ENERGY SOLUTIONS, http://www.c2es.org/us-states-regions/regional-climate-initiatives/rggi.

[178] *What's Being Done in the U.S.*, CTR. FOR CLIMATE & ENERGY SOLUTIONS, http://entergy.c2es.org/climate-change/what-is-being-done-in-the-us.

[179] *Id.*

review of the RGGI program, the RGGI states amended the RGGI model rule (in February 2013) to substantially reduce the emissions cap from 165 mtCO$_2$ [metric tons] to 91 mtCO$_2$."[180] The Western Climate Initiative among California, British Columbia, and Quebec and the Midwest Greenhouse Gas Reduction Accord are additional regional efforts by governors of those states to reduce GHG emission in the electricity and transportation sectors.[181]

VI. Environmental Regulation of Hydropower

One low-carbon means of producing electricity, which is also controversial due to its environmental and social effects beyond the realm of climate change, is hydroelectric power. "Conventional" hydropower produced 6–9 percent of U.S. electricity generation each year between 1998 and 2014, representing more than half of the nation's renewable electricity generation.[182] Hydropower uses several different technologies, as the Congressional Research Service explains:

> Conventional hydropower plants take three general forms: storage (or impoundment), run-of-the river (or diversion), and pumped storage. A storage plant uses a dam to store enough water in a reservoir so that, when released, it flows through a penstock to a turbine, spinning it, which in turn activates a generator to produce electricity. A run-of-river plant directs a portion of a river through a canal or penstock to generate electricity without the need for a reservoir. A pumped storage facility stores energy and generates electricity by pumping water from a lower reservoir to an upper reservoir during off-peak hours, releasing the stored water from the upper reservoir to the lower reservoir during periods of higher electricity demand.[183]

Three western states—Washington, California and Oregon— have "more than half of the total U.S. hydroelectric capacity for electricity generation."[184] The federal government, through the Bureau of Reclamation, Army Corps of Engineers, and Tennessee Valley Authority, operates 8 percent of the country's hydroelectric facilities but approximately 50 percent of hydropower generation.[185]

---

[180] JONATHAN L. RAMSEUR, CONG. RESEARCH SERV., THE REGIONAL GREENHOUSE GAS INITIATIVE: LESSONS LEARNED AND ISSUES FOR CONGRESS (Apr. 27, 2016).

[181] *See Multi-State Climate Initiatives*, CTR. FOR CLIMATE & ENERGY SOLUTIONS, http://www.c2es.org/us-states-regions/regional-climate-initiatives.

[182] KELSI BRACMORT, ET AL., CONG. RESEARCH SERV., HYDROPOWER: FEDERAL AND NONFEDERAL INVESTMENT 2–3 (July 7, 2015).

[183] *Id.*

[184] *Hydropower*, CTR. FOR CLIMATE & ENERGY SOLUTIONS, http://www.c2es.org/technology/factsheet/hydropower.

[185] *See* U.S. DEP'T OF ENERGY, 2014 HYDROPOWER MARKET REPORT v (Apr. 2015).

The other 92 percent of U.S. hydroelectric facilities are operated by the private sector, public utilities, and state or local governments, and produce the other 50 percent of U.S. hydropower generation.[186] The public sector (federal agencies, states, cooperatives, and municipally-owned utilities) own and operate the larger hydropower plants—those greater than 30 megawatts—and thus generate approximately 90 percent of hydropower capacity. The private sector, in turn, (public utilities, private nonutility, and industrial entities) owns over 80 percent of the hydropower plants (which tend to be "small hydro and lower power plants") and generates about 10 percent of U.S. hydropower capacity.[187]

The nonfederal hydropower projects are regulated by FERC under the Federal Power Act (FPA) and range from large dams and reservoirs occupying tens of thousands of acres, down to "microhydro" projects.[188] FERC has jurisdiction over hydropower plants that (1) are located on navigable waters of the United States; (2) occupy "public lands or reservations of the United States"; or (3) "utilize surplus water or water power from any [U.S.] Government dam."[189] As a practical matter, nearly all hydropower projects in the United States other than the projects owned and operated by federal agencies are regulated by FERC under the FPA. There are exceptions, although Congress has directed FERC to consider exercising authority over facilities that currently do not fall within FERC jurisdiction while expressly exempting other facilities from this jurisdiction. As FERC explains, in 2013, Congress enacted the Hydropower Regulatory Efficiency Act of 2013, which, among other things "exempts certain conduit hydropower facilities from the licensing requirements" of the FPA, "defines 'small hydroelectric power projects' as having an installed capacity that does not exceed 10,000 kilowatts," and "directs the Commission to investigate the feasibility of a 2-year licensing process for hydropower development at non-powered dams and closed-loop pump storage projects."[190]

With respect to the regulation of federal hydropower projects, FERC does not regulate or license these facilities under the FPA. Instead, these facilities are governed in accordance with regulations and manuals adopted pursuant to the congressional legislation

---

[186] *Id.* at v-vi.

[187] BRACMORT, *supra* note 182, at 4–5.

[188] For a detailed explanation of federal regulation of hydropower facilities, see Charles R. Sensiba, *Hydropower, in* THE LAW OF CLEAN ENERGY: EFFICIENCY AND RENEWABLES (Michael B. Gerrard, ed. 2011).

[189] 16 U.S.C. § 817(1) (West 2016).

[190] Pub. L. No. 113-23 (Aug. 9, 2013).

authorizing the specific facility and the operating entities' statutory authority.

For facilities under FERC jurisdiction, the first step in obtaining authorization to construct a dam is the issuance of a preliminary permit, which determines which applicants will be first in line to receive a hydropower license and creates a period—up to three years—in which environmental and other studies may be conducted.[191] Following this preliminary period, FERC—if it determines that the facility will meet the requirements discussed directly below—may issue a license for the operator to construct and operate a hydropower facility "for a period not exceeding 50 years,"[192] and relicensed for thirty to fifty years.[193]

Hydropower licensing requires that entities proposing to construct and operate dams meet a variety of environmental protection provisions. The entity proposing the license must provide information that shows that the project will be:

> [b]est adapted to a comprehensive plan for improving or developing a waterway or waterways for the use or benefit of interstate or foreign commerce, for the improvement and utilization of water-power development, for the adequate protection, mitigation, and enhancement of fish and wildlife (including related spawning grounds and habitat), and for other beneficial public uses, including irrigation, flood control, water supply, and recreational and other purposes. . . .[194]

In considering whether to issue the license, FERC must "give equal consideration" to many of the same factors that must be provided in the information described above, including consideration of "the purposes of energy conservation, the protection, mitigation of damage to, and enhancement of, fish and wildlife (including related spawning grounds and habitat), the protection of recreational opportunities, and the preservation of other aspects of environmental quality."[195] If FERC grants the license, the license must contain conditions for the "protection, mitigation, and enhancement" of fish and wildlife.[196] Further, if the Secretaries of the Interior or Commerce require it, the licensee must construct "fishways" that allow fish to safely move through or around the hydro facility.[197]

---

[191] 16 U.S.C. § 798 (West 2016).
[192] 16 U.S.C. § 799 (West 2016).
[193] 16 U.S.C. § 808(e) (West 2016).
[194] 16 U.S.C. § 803(a) (West 2016).
[195] 16 U.S.C. § 797(e) (West 2016).
[196] 16 U.S.C. § 803(j) (West 2016).
[197] 16 U.S.C. § 811 (West 2016).

Finally, before issuing the license FERC must complete an environmental review of the impacts of the project under the National Environmental Policy Act.[198] If any historic or cultural resources will be impacted FERC must consult with state historic preservation officers and other parties under the National Historic Preservation Act to discuss possible means of preventing or mitigating impacts on these resources, and it must consult with the Fish and Wildlife Service regarding impacts on endangered and threatened species and, depending on the species present, obtain permits under the Endangered Species Act.[199] Further, for projects near the coast FERC must ensure that its actions will be consistent with state Coastal Zone Management Plans—plans designed to protect coastal and near-coastal resources from the impacts of federally-approved projects, among other projects. And for all projects licensed, affected states must certify that the project will not cause a violation of the state's standards for water quality as established under the CWA § 401.[200]

Water quality certification under CWA § 401 confers significant and broad authority to the states in the FERC licensing process. Section 401(d) provides:

> Any certification provided under this section shall set forth any effluent limitations and other limitations, and monitoring requirements necessary to assure that any applicant for a Federal license or permit will comply with any applicable effluent limitations and other limitations, ... standard of performance ..., or prohibition, effluent standard, or pre-treatment standard ..., and with any other appropriate requirement of State law set forth in such certification, and shall become a condition on any Federal license or permit subject to the provisions of this section.

Thus, in issuing Section 401 certifications, the states are authorized to impose conditions they consider necessary to ensure compliance with state water quality standards, as well as any other "appropriate requirement of State law," which become conditions of the FERC-issued license.[201] FERC has no authority to reject or modify Section 401 conditions, and if a state denies certification, FERC will not issue the license.

---

[198] HANDBOOK FOR HYDROELECTRIC PROJECT LICENSING AND 5 MW EXEMPTIONS FROM LICENSING, FED. ENERGY REGULATORY COMM'N B–1 (Apr. 2004).

[199] *Id.* at B–2.

[200] *Id.* at D–3.

[201] 33 U.S.C. § 1341(d) (West 2016).

For decades, there have been heated disagreements over whether hydropower should count as "renewable" or "clean" energy. It is a replenishable resource (at least in times of sufficient water resources) that is not derived from fossil fuels, and it does not emit GHGs when generating electricity. But it has been subject to significant criticism for its environmental impacts on ecosystems, fish, water-based birds, and recreation, as well as concerns about declining water resources, particularly in western states.

VII. Regulation of Nuclear Power

Another controversial, zero-carbon source of energy is nuclear power. Nuclear power plants use uranium pellets, which are packed into long, thin rods called "fuel rods," to generate a chain reaction. The chain reaction occurs when the relatively unstable uranium atoms break, releasing neutrons. These neutrons then hit other uranium atoms, causing other nuclei of the atoms to break, thus releasing more neutrons. The breaking of the nuclei produces heat that is essential to electricity generation: this heats up water, produces steam, and turns a turbine to produce electricity. The fuel rods in which the chain reaction occurs are packed within a device called a "reactor core," which can be lowered or raised into cool water in order to speed up or slow the chain reaction. There are two methods by which the heat from the reaction turns a turbine. In the more common pressurized water reactor, the heated water from inside the reactor core becomes very hot but does not boil because it is kept under pressure.[202] This water is circulated near a separate set of pipes carrying water that is not in contact with the reactor. The heat from the water near the reactor core passes through a heat exchanger to the water in the separate set of pipes, and the water in the separate set of pipes produces steam, which turns a turbine.[203] In the less common boiling water reactor, the water near the reactor core boils and produces steam, which directly turns a turbine.[204]

The most important law originally governing nuclear power in the United States, as well as military uses of nuclear material, was the Atomic Energy Act of 1954, which provided that "the development, use, and control of atomic energy shall be directed so as to make the maximum contribution to the general welfare" and to "promote world peace."[205] The Act initially gave one agency—the Atomic Energy Commission—authority over both civilian and military uses of nuclear material, but the Energy Reorganization Act of 1974 gave the Department of Energy jurisdiction over nuclear

---

[202] PAUL BREEZE, POWER GENERATION TECHNOLOGIES (2d. ed. 2014).
[203] *Id.*
[204] *Id.*
[205] 42 U.S.C. § 2011 (West 2016).

weapons, and it gave the Nuclear Regulatory Commission (NRC) jurisdiction over nuclear power plants and their safety as well as many aspects of nuclear waste disposal.[206]

The permitting of nuclear power plants by the NRC is a complex process. Utilities or other entities wishing to construct a plant can select a two-step licensing process, in which they obtain a permit to construct the plant and a separate permit to operate the plant.[207] Alternatively, developers of nuclear power plants can obtain: (1) an early site permit (ESP), which designates and "holds" a potential site for a project but does not commit the company to the project, (2) certification of plant design, which is a type of "pre-approval" of a particular design and again, does not commit the company to the project,[208] and (3) a combined construction and operating license. As the NRC explains, in the two-step licensing process the applicant must first submit "preliminary safety analyses," an "environmental review," and "financial and antitrust statements" and an explanation of why the plan is needed when it applies for a construction permit.[209] The NRC prepares a "safety evaluation report" based on this information, including "findings on site safety characteristics and emergency planning," as well as an environmental impact statement under the National Environmental Policy Act.[210] Further, the Advisory Committee on Reactor Safeguards (ACRS) reviews the application and safety evaluation report and "reports its results" to the NRC, and the Atomic Safety Licensing Board holds a public hearing on the application. Only after these and other steps are complete may the NRC issue a construction permit.[211] To obtain an operating license, the developer of the plant must, in addition to an application for the license, submit a "final safety analysis report and an updated environmental report," including information on "the plant's final design, safety evaluation, operational limits, anticipated response of the plant to postulated accidents, and plans for coping with emergencies."[212] The Federal Emergency Management Agency reviews parts of the application, as does the ACRS,[213] before the NRC may issue this second license.

---

[206] *Governing Legislation*, NUCLEAR REGULATORY COMM'N, http://www.nrc.gov/about-nrc/governing-laws.html.

[207] NUCLEAR REGULATORY COMM'N, NUCLEAR POWER PLANT LICENSING PROCESS 1 (2004), http://www.nrc.gov/reading-rm/doc-collections/nuregs/brochures/br0298/br0298r2.pdf.

[208] *Id.*

[209] *Id.* at 3.

[210] *Id.*

[211] *Id.*

[212] *Id.* at 4.

[213] *Id.*

In the alternative licensing process, the developer of a proposed plant may apply separately for an early site permit just to designate a site for the plant. The NRC can approve this permit to last for at least ten years to a maximum period of twenty years, and the permit is renewable for a ten or twenty-year period.[214] The developer also can apply for a standard design certification that lasts fifteen years and is renewable for ten to fifteen years.[215] Finally, the combined construction and operating permit introduced above lasts forty years and is renewable for a twenty-year period.[216]

The NRC is the sole entity that may approve or reject a nuclear power plant on the basis of safety. Because the Atomic Energy Act gave the Atomic Energy Commission jurisdiction over all aspects of nuclear materials, including—as summarized by the Ninth Circuit—authority over the "license the transfer, delivery, receipt, acquisition, possession and use of nuclear materials,"[217] courts have interpreted the Act to have occupied the entire field of the regulation of nuclear safety.[218] Thus, states may not regulate the safety aspects of the design or construction of power plants or their operation due to federal preemption. However, as discussed in Chapter 6, states retain jurisdiction over the decision of whether a utility should build a power plant or not. Thus, states may still deny the construction of nuclear power plants altogether on the basis that these plants are not "needed" or will result in unjust or unreasonable rates. So when California decided in the 1970s that a nuclear power plant would not be approved in the state until a centralized, long-term nuclear waste disposal facility was available, citing to concerns about the costs of alternative methods of disposal that could be passed on to consumers of electricity, the U.S. Supreme Court determined that this state law was not preempted.[219] The Court emphasized that the "[n]eed for new power facilities, their economic feasibility, and rates and services, are areas that have been characteristically governed by the States."[220]

Still, states may not use economic considerations as a mere pretense for prohibiting the construction of nuclear plants. Thus, when the Vermont Legislature decided that a nuclear plant in the state could not continue operating without the legislature's approval,

---

[214] *Id.* at 6; 10 C.F.R. §§ 52.26, 52.33 (West 2016).

[215] NUCLEAR REGULATORY COMM'N, *supra* note 207, at 8; 10 C.F.R. §§ 52.55, 52.61 (West 2016).

[216] NUCLEAR REGULATORY COMM'N, *supra* note 207, at 9; 10 C.F.R. § 50.51; 10 C.F.R. § 54.31 (West 2016).

[217] Pac. Gas & Elec. Co. v. State Energy Res. Conservation & Dev. Comm'n, 461 U.S. 190, 207 (1983).

[218] Entergy Nuclear Vermont Yankee v. Shumlin, 733 F.3d 393, 409 (2d Cir. 2013).

[219] *Pac. Gas & Elec.*, 461 U.S. at 198, 216.

[220] *Id.* at 205.

Sec. VII REGULATION OF NUCLEAR POWER 137

asserting that it wanted to encourage alternative, renewable sources of energy and cheaper energy options, the Second Circuit Court of Appeals was not convinced.[221] Vermont lies within an area in which transmission lines are operated by a regional organization, and the court noted that these entities—Independent System Operators and Regional Transmission Organizations—run competitive electricity markets that allow for the flourishing of independent energy generation of all types, including renewable sources and other types of generation that may be cheaper than nuclear power.[222] The state therefore did not need to restrict nuclear power in order to give these other sources of power a chance to succeed in the marketplace according to the court. Further, legislators' statements about safety concerns in discussions leading up to the passage of the bill[223] persuaded the court that federal law preempted Vermont's legislation.[224]

California's decision in the 1970s to place a moratorium on the approval of new nuclear power plants within the state until a long-term nuclear waste disposal facility was available has essentially transformed into a ban on nuclear plants in that state; there seem to be few prospects of such a facility being approved. The regulation of nuclear waste disposal is as complex—or perhaps even more so—than the regulation of nuclear power plants themselves, and decisions about long-term disposal options are particularly thorny. There are three types of radioactive waste—low-level waste such as wastes associated with hospitals' radiological units, mill tailings from uranium mining, and high-level wastes. Low-level radioactive waste disposal is regulated by the NRC and the states, which can enter into compacts by which they agree upon the siting of disposal facilities and the states from which wastes will be accepted at these disposal facilities.[225] The NRC also regulates the handling and disposal of mill tailings,[226] but high-level nuclear waste disposal involves more agencies. The DOE must build and operate the repository for high-level waste if a site for this facility is ever agreed upon; the EPA must regulate the environmental safety aspects of the facility, including potential human exposure to radiation from the facility over millions of years; and the NRC must license the

---

[221] *Entergy Nuclear Vermont Yankee*, 733 F.3d at 403.

[222] *Id.* at 411–14, 416.

[223] *Id.* at 420–21.

[224] *Id.* at 427.

[225] *Low-Level Waste Disposal*, U.S. NUCLEAR REGULATORY COMM'N, http://www.nrc.gov/waste/llw-disposal.html.

[226] *Uranium Mill Tailings*, U.S. NUCLEAR REGULATORY COMM'N, http://www.nrc.gov/waste/mill-tailings.html.

repository by granting a license to the DOE to construct and operate the facility.

If the repository is ever built, it will be funded through payments made by nuclear power plant operators into the Nuclear Waste Fund.[227] Nuclear power plant operators already pay into this fund through a fee charged per unit of electricity that they generate, and many have requested and obtained damages because they have contributed to the fund, yet they lack a repository to which they can send their waste.

The most likely location for a centralized, high-level nuclear waste repository is Yucca Mountain in Nevada, which has been considered as a site since the 1980s. After Congress mandated that the EPA develop environmental standards for the Yucca Mountain repository (which the EPA did), and directed the DOE to characterize Yucca Mountain and determine its suitability (which the DOE did), Congress approved the site as suitable for high-level waste disposal. The DOE in 2008 applied to the NRC for a license for the Yucca Mountain facility. By that time, however, President Obama had made campaign pledges to oppose Yucca Mountain, and the NRC "shut down its review and consideration of the Department of Energy's license application."[228] After several states sued the NRC, the D.C. Circuit issued a "writ of mandamus against the Commission," holding that "unless and until Congress authoritatively says otherwise or there are no appropriated funds remaining, the Nuclear Regulatory Commission must promptly continue with the legally mandated licensing process."[229] The NRC had tried to argue that inadequate funds kept it from properly considering the application, but the court noted that the NRC had *some* funds and could not automatically assume that it would not have funding in the future.[230] Further, the court rejected other arguments of the NRC, such as the argument that it could use executive prosecutorial discretion to avoid considering the license.[231] Although the court found that the NRC could not use a lack of funding to refuse to consider the application, a continued dearth of funds has allowed the NRC to drag its feet, and few serious steps toward considering the license have occurred. Thus, the U.S. still lacks a centralized high-level nuclear waste facility, and this situation is unlikely to change in the near future.

---

[227] 42 U.S.C. § 1022 (West 2016).
[228] *In re Aiken Cty.*, 725 F.3d 255, 258 (D.C. Cir. 2013).
[229] *Id.* at 267.
[230] *Id.* at 260.
[231] *Id.* at 263.

Chapter 5

ENERGY FINANCING, INCENTIVES, AND MANDATES

I. Introduction

Although many contend that the United States lacks a cohesive energy policy, the federal government has encouraged and discouraged various types of energy development over time, primarily through acts that add to and amend the Internal Revenue Code.[1] Federal agencies also have spurred various types of energy development by mapping the locations of fossil fuel and renewable resources[2] directly and in partnership with industry, investing in research and development for emerging energy technologies,[3] and providing loan support and other incentives for energy development,[4] among other actions.[5] Historically, federal support for energy development focused primarily on fossil fuels. More recently the federal government has directed the majority of energy incentives toward renewable energy, but these incentives will soon expire.[6]

Because of the predominance of energy policy within the federal tax code, the first section of this Chapter focuses on tax incentives for energy production and describes the historic and current tax incentives for both fossil fuels and renewable energy. The second section focuses on present-day government mandates for the purchase and use of certain types of renewable energy. These include state renewable portfolio standards (RPSs), feed-in tariffs (FiTs), net metering in the electricity sector, and the federal Renewable Fuel Standard (RFS) in the transportation sector. Finally, the third section briefly explores creative private financing mechanisms allowed by federal regulation as well as traditional private financing

---

[1] *See* Linda Larson & Dustin Till, *The Right Fit to Promote Renewable Energy?*, 40 No. 2 ABA TRENDS 10 (2008) (noting that one of several "key questions for policymakers" is "whether to continue to rely on the federal tax code to set energy policy"); MARK BOLINGER, LAWRENCE BERKELEY NATIONAL LABORATORY, AN ANALYSIS OF THE COSTS, BENEFITS, AND IMPLICATIONS OF DIFFERENT APPROACHES TO CAPTURING THE VALUE OF RENEWABLE ENERGY TAX INCENTIVES i (May 2014) ("In the United States, Federal incentives for the deployment of wind and solar power projects are delivered primarily through the tax code. . . .").

[2] John M. Golden & Hannah J. Wiseman, *The Fracking Revolution: Shale Gas as a Case Study in Innovation Policy*, 64 EMORY L. J. 955, 991, 973 n. 97, 1032 (2015).

[3] *Id.* at 983–89; Wind Research and Development, Dep't of Energy, http://energy.gov/eere/wind/wind-research-and-development.

[4] Pub. L. No. 114–113.

[5] *See* MOLLY F. SHERLOCK & JEFFREY M. STUPAK, CONG. RESEARCH SERV., ENERGY TAX INCENTIVES: MEASURING VALUE ACROSS DIFFERENT TYPES OF ENERGY RESOURCES 1 (Mar. 19, 2015), https://www.fas.org/sgp/crs/misc/R41953.pdf (describing various types of incentives, with an emphasis on the tax code).

[6] *Id.*

devices such as power purchase agreements between renewable energy developers and wholesale buyers of electricity.

These materials illustrate the importance of government incentives and mandates in influencing the production and use of energy resources both at the time the incentives were put in place and also, in the case of fossil fuels, long after their expiration because of the security such long-term support for an industry provides. They also demonstrate the growing set of private financing mechanisms available to developers. This Chapter does not address the numerous other forms of government support for the energy sector, such as research grants and direct investments in research and development; coal, oil, and gas leases on federal lands; or laws limiting liability for nuclear or other energy facilities.

II. Tax Incentives for Energy Production and Use

For nearly a century, Congress has provided "tax expenditures" to encourage energy production and use. Tax expenditures are defined as "federal revenue losses"[7] that arise from tax provisions granting special exemptions, deductions, income deferments, or reduced rates on income that would otherwise be taxable. Of course, tax expenditures are only one way to encourage certain types of energy production and use by making them more profitable. Other options are to tax energy uses the government wants to discourage, such as placing a carbon tax on industries to discourage the use of fossil fuels. Yet another alternative is to mandate the use of certain types of energy or create minimum prices for preferred energy resources (discussed later in this Chapter). All of these tools are options to attempt to correct market failures associated with the production and use of energy, including the positive and negative environmental externalities associated with various energy resources and the need for R & D support and other incentives for emerging technologies in the energy sector.

The total estimated tax expenditures associated with the energy sector were approximately $20 billion in 2010, $24.2 billion in 2012, $23.3 billion in 2013, and $16.7 billion in 2014.[8] As Molly Sherlock and Jeffrey Stupak explain, the primary cause of the drop between 2012 and 2014 was the expiration of the ethanol tax credits at the end of 2011,[9] the expiration that same year of the Section 1603 grants in lieu of tax credits program for certain renewable energy projects,

---

[7] *Id.*

[8] *Id.* at 7–8 & Tbl. 2.

[9] *Expired, Repealed, and Archived Incentives and Laws*, U.S. DEP'T OF ENERGY, ALTERNATIVE FUELS DATA CTR., http://www.afdc.energy.gov/laws/laws_expired?jurisdiction=US (describing the expiration of the "Small Ethanol Producer Tax Credit" and the "Volumetric Ethanol Excise Tax Credit (VEETC)," which provided tax credits to ethanol blenders).

and the expiration of certain other renewable energy tax incentives since 2011[10] (all of which are discussed later in this Chapter). Until 2012, when the primary tax incentive for ethanol expired, more than half of tax benefits for renewable energy went to the biofuels industry.[11] Overall, with regard to the $16.7 billion in energy tax incentives in 2014, $4.5 billion (26.8%) went to fossil fuel production, and $8.8 billion (52.8%) went to renewable energy production (including biofuels).[12] The remainder went primarily to nuclear energy and energy efficiency programs.[13] This stands in sharp contrast to the years prior to 2005, when virtually all tax expenditures and benefits went to oil and gas and other fossil fuel production with a negligible amount directed to renewable energy resources.

But these percentages are expected to change dramatically. The fossil fuel tax incentives are permanent parts of the tax code that have been in place for decades and do not require Congressional action to continue.[14] By contrast, the renewable energy tax incentives were created with expiration dates, which means Congress must act for these incentives to continue. Over the past two decades, these incentives have expired or have almost expired numerous times and will likely expire permanently by approximately 2020. The on and off nature of these tax incentives creates investment uncertainty in the renewable energy sector as well as "boom and bust" cycles for renewable energy development.[15] Sherlock and Stupack note that due to these expirations, a "substantial shift in balance of energy tax incentives across different types of energy resources in projected to occur."[16] Absent Congressional action to reduce oil and gas tax incentives, this industry may return to its original dominant position with regard to tax incentives as compared to the renewable energy sector.

A. Fossil Fuel Tax Credits and Incentives

The U.S. government first supported fossil fuels in the early 1900s through provisions that allowed for generous deductions of expenses—including intangible drilling costs (such as labor and tool rentals), the costs of drilling dry holes (non-producing wells), and a "percentage depletion allowance" through which producers deducted

---

[10] SHERLOCK & STUPAK, *supra* note 5, at 6.

[11] *Id.* at 6–7 & Tbl. 2.

[12] *Id.* at 8.

[13] *Id.*

[14] *See, e.g.*, TREASURY.GOV., UNITED STATES—PROGRESS REPORT ON FOSSIL FUEL SUBSIDIES, PART 1: IDENTIFICATION AND ANALYSIS OF FOSSIL FUEL PROVISIONS (2014).

[15] Felix Mormann, *Requirements for Renewables Revolution*, 38 ECOL. L.Q. 903, 950 (2011).

[16] SHERLOCK & STUPAK, *supra* note 5, at 6.

"from their gross receipts"[17] a percentage of the revenue generated by oil wells prior to paying taxes.[18] These benefits had a cumulatively large effect: "From 1977 until the Mid-1980s, tax benefits associated with these incentives ranged from $5 billion to over $10 billion in inflation adjusted dollars."[19] These benefits also remain in place after a temporary repeal in the 1980s.[20] The current percentage depletion allowance provides that oil and gas producers may annually deduct from the taxable revenue fifteen percent of the first 1,000 barrels of oil that they produce daily,[21] and it saved oil and gas companies $900 million in the year 2011 alone.[22] Additional tax benefits for oil and gas development include, *inter alia*, rapid amortization of the costs of identifying oil and gas formations and exploring their likely productivity, and deduction of a percentage of oil and gas companies' taxable income.[23] These tax incentives remain permanent parts of the Tax Code unless and until Congress affirmatively repeals them. Congress provided additional tax benefits in the 1990s and 2000s to stimulate unconventional production of oil and gas. Coal bed methane gas producers and hydraulic fracturing producers captured the bulk of these tax benefits, which were ultimately phased out in the Energy Policy Act of 2005.[24]

As of 2014, annual tax expenditures for fossil fuels totaled $4.5 billion.[25] These consisted of credits for investments in clean coal facilities, amortization of air pollution control equipment for coal-fired electric plants, the oil and gas development tax benefits described above, "15-year depreciation for natural gas distribution lines," an option to "expense 50% of qualified refinery costs," "exceptions for publicly traded partnerships with qualified income from energy related activities," and alternative fuel credits.[26]

---

[17] Golden & Wiseman, *supra* note 2, at 991 (quoting John A. Bogdanski, *Reflections on the Environmental Impacts of Federal Tax Subsidies for Oil, Gas, and Timber Production*, 15 LEWIS & CLARK L. REV. 323, 325 (2011)).

[18] Alexandra B. Klass, *Tax Benefits, Property Rights, and Mandates: Considering the Future of Government Support for Renewable Energy*, 20 J. ENVTL. & SUSTAINABILITY L. 19, 31–32 (2013); Golden & Wiseman, *supra* note 2, at 991 (describing the allowance).

[19] Klass, *supra* note 18, at 32.

[20] *Id.*

[21] *Id.*

[22] Golden & Wiseman, *supra* note 2, at 955 (citing Cong. Budget Office, Federal Financial Support for the Development and Production of Fuels and Energy Policies 2–3 (2012)).

[23] *Id.* at 992–94; Calvin H. Johnson, *Accurate and Honest Tax Accounting for Oil and Gas*, 125 TAX NOTES 573, 577 (2009) (describing the major tax benefits for oil and gas).

[24] Klass, *supra* note 18, at 31–32.

[25] SHERLOCK & STUPAK, *supra* note 5, at 6–7 & Tbl. 2.

[26] *Id.* at 7, Tbl. 2; MOLLY F. SHERLOCK, ENERGY TAX POLICY: ISSUES IN THE 113TH CONGRESS Tbl. 1 (Dec. 19, 2013).

B. Major Renewable Energy Tax Credits and Incentives

As noted above, as of 2014, over 50 percent of the total tax expenditures for the energy industry went to renewable energy, although that percentage is expected to decrease significantly in a few years as a result of the expiration of many of these provisions. The bulk of the renewable energy tax credits and incentives Congress has put in place since the 1990s have promoted: (1) renewable sources of electricity (primarily wind and solar energy) and (2) biofuels.

Policymakers have recognized that the long history of government financial support for fossil fuels made it difficult for developing wind, solar, and other renewable energy resources to compete on cost and availability. But rather than eliminate the longstanding tax support for fossil fuels, Congress chose instead to create new tax incentives for renewable energy resources. However, unlike the tax incentives for fossil fuels, which were created as permanent provisions of the Tax Code, the tax incentives for renewable energy were designed to be temporary with set expiration dates, and thus require additional action by Congress as those dates approach to extend those benefits. As introduced above, this structure has resulted in a number of "boom and bust" cycles for wind and solar energy along with significant investment uncertainty.[27] There remains significant debate over whether renewable resources are sufficiently "established" to stand on their own without government support, or whether the long history of support for fossil fuel production, which continues today, warrants continued government support for renewable energy resources. The primary tax incentives for renewable energy are detailed below, including some that have expired but that have had a lasting effect on the industry.

Section 45 Production Tax Credit (PTC): In the Energy Policy Act of 1992, Congress created a PTC (currently 2.3 cents per kWh) starting in 1994 for wind and closed-loop biomass energy sold to an unrelated party for the first 10 years of the project's operation. The PTC was later expanded to include the production of several other renewable energy resources such as solid waste, geothermal energy, qualified hydropower, and hydrokinetic facilities,[28] although only wind, geothermal, and closed-loop biomass receive the full 2.3 cents per kWh credit while the others receive 1.1 cents per kWh.[29]

---

[27] Mormann, *supra* note 15, at 918.

[28] MOLLY F. SHERLOCK, CONG. RESEARCH SERV., THE RENEWABLE ELECTRICITY PRODUCTION TAX CREDIT: IN BRIEF 4 (July 14, 2015).

[29] *Id.*

These credit amounts change annually to account for inflation.[30] The PTC has never been granted to solar projects.[31]

The PTC is based on production of electricity rather than the cost of or investment in the project itself.[32] It therefore incentivizes projects not only to be constructed, but also to begin operating as soon as possible so that the owner of the renewable energy infrastructure can receive tax benefits sooner. The PTC is a major driver for wind energy development, allowing wind to be more cost-competitive with fossil-fuel electricity resources. The PTC reduces or eliminates the amount of income tax a project owes but is only valuable if there are taxes owed in the first place.[33] In many cases, wind energy developers cannot take full advantage of the PTC because there are other tax incentives, such as "accelerated depreciation," which makes the value of companies' "long-lived assets" appear lower for tax purposes and can cause companies to report net operating losses on their tax returns.[34] This has caused scholars and policy experts to advocate for alternative ways to allow the tax incentive to be used, such as allowing the deductions to be applied against income of the renewable energy developer that is unrelated to the renewable energy project; carrying the tax benefits forward, beyond the point when a company reports net operating losses; or allowing a third-party tax equity investor with higher tax liability to invest in the project for so long as the company can benefit from the PTC, thus providing needed capital to the renewable energy project and receiving tax benefits.[35] We discuss these options in more detail in the "energy financing" portion of this Chapter.

Because it is not a permanent part of the Tax Code like the tax incentives for fossil fuels, the PTC is often subject to expiration. It expired nine times between its enactment and 2013, when Congress renewed it as part of the "fiscal cliff" budget negotiations that year.[36] In 2013, Congress (with IRS clarification) made the deadline tied to the construction date of the project rather than the in-service date of the project, thus allowing projects to claim the credit so long as they started construction by the deadline.[37] In 2015, the PTC was again set to expire but Congress extended it an additional time (as

---

[30] *Id.*

[31] BOLINGER, *supra* note 1, at 6.

[32] *Id.* at 8.

[33] *Id.* at 7.

[34] *Id.* at 5.

[35] *See id.* at 6–10; Felix Mormann, *Beyond Tax Credits: Smarter Tax Policy for a Cleaner, More Democratic Energy Future*, 31 YALE J. ON REG. 303 (2014).

[36] BOLINGER, *supra* note 1, at 6–7.

[37] *Id.* at 38.

described below) in exchange for a lifting of the decades-old ban on U.S. crude oil exports.

Section 48 Investment Tax Credit (ITC): The ITC is the primary tax incentive for solar projects. The first ITCs for renewable energy date back to 1978. Unlike the PTC, which is based on the production of electricity, the ITC is tied to the amount of investment in a qualifying project.[38] The Energy Policy Act of 2005 temporarily increased the ITC from 10 percent to 30 percent of the project investment cost beginning in 2006, and Congress extended the 30 percent level again in 2008 for projects begun by the end of 2016, "at which point it [was] scheduled to revert back to 10%."[39] As explained below, however, Congress extended the 30 percent level again in December 2015 to projects commencing construction in or before 2020. The ITC applies to a wide variety of renewable energy projects, including solar, combined heat and power, fuel cells, microturbines, geothermal, and small wind projects. PTC-eligible projects also may opt for the ITC in lieu of the PTC. For a developer that takes advantage of the ITC, the credit is realized in full the same year that a project begins commercial operations but "vests linearly over a 5-year period," which means that if the project is sold within those five years, the Internal Revenue Service takes back the credit.[40]

Section 1603 Cash Grant Program: The American Recovery and Reinvestment Act of 2009 (ARRA) provided ITC and PTC-qualified facilities the option of receiving up to 30 percent of the costs of a renewable energy project in the form of non-taxable direct cash payment instead of claiming either credit,[41] which allowed the immediate benefit of the funds even if the company might not have otherwise been able to use it against its tax liability.[42] As summarized by Mark Bolinger of the Lawrence Berkeley National Laboratory, different types of renewable energy projects had different requirements for when the project had to be commenced and/or placed in service to qualify for the cash grant. Congress created the program as a "temporary response to a severe shortage of tax equity investors"[43]—large companies with relatively high tax liability, which provided funds to cash-strapped renewable developers in exchange for receiving tax benefits of the renewable

---

[38] *Id.* at 8.
[39] *Id.*
[40] *Id.*
[41] *Id.*
[42] PHILLIP BROWN & MOLLY F. SHERLOCK, CONG. RESEARCH SERV., ARRA SECTION 1603 GRANTS IN LIEU OF TAX CREDITS FOR RENEWABLE ENERGY: OVERVIEW, ANALYSIS, AND POLICY OPTIONS 2 (Nov. 9, 2011).
[43] BOLINGER, *supra* note 1, at 8.

energy project—after the financial crisis in 2008 and 2009.[44] Projects eligible to elect the grant had to be under construction by 2010 (later extended to 2011) and placed in service by the end of 2012 (for wind projects) or 2016 (for solar projects).[45] As a result, expenditures under the program have continued since 2011 but have begun to decline. The program outlays were $8.1 billion in 2013 and $24 billion over the 2010–2014 time period.[46]

Sections 1703 and 1705 U.S. Department of Energy Loan Guarantee Program: In addition to tax credits administered by the IRS, the Department of Energy has a loan guarantee program originally created through the Energy Policy Act of 2005 and expanded through ARRA. Section 1703 of the program, which remains available for applications through July 2016,[47] applies to emerging commercial alternative energy and carbon capture and sequestration projects.[48] Section 1705 of the Energy Policy Act of 2015—which only applied to projects under construction by September 30, 2011—supported more traditional, established renewable energy projects like solar and wind as well as transmission projects and biofuels.[49] As of 2014 the DOE had provided "2 section 1703 loan guarantees totaling $6.2 billion in support of a nuclear power generation project" and "27 section 1705 loan guarantees totaling $13.6 billion in support of 24 energy projects including solar and wind generation, geothermal and biomass generation, solar panel manufacturing, energy storage, and electricity transmission."[50]

Clean Renewable Energy Bonds (CREBs): In addition to providing loans for renewable energy projects, the federal government provides money for public entities like state and local governments, Indian tribes, and consumer-owned utilities (electric cooperatives) to issue bonds to finance renewable energy projects.[51] As the Department of Energy explains, the bonds are typically for projects that are PTC-eligible. Congress provides the funds for the

---

[44] *Id.* at 8; Mormann, *supra* note 35, at 313.

[45] BROWN & SHERLOCK, *supra* note 42, at 25–27; SHERLOCK & STUPAK, *supra* note 5, at 15.

[46] SHERLOCK & STUPAK, *supra* note 5, at 15.

[47] *Program Info*, U.S. DEP'T OF ENERGY-LOAN GUARANTEE PROGRAM, http://energy.gov/savings/us-department-energy-loan-guarantee-program.

[48] *Section 1703 Loan Program*, U.S. DEP'T OF ENERGY, http://energy.gov/lpo/services/section-1703-loan-program.

[49] *Section 1705 Loan Program*, U.S. DEP'T OF ENERGY, LOAN PROGRAMS OFFICE, http://energy.gov/lpo/services/section-1705-loan-program.

[50] U.S. GOV'T ACCOUNTABILITY OFFICE, DOE LOAN PROGRAMS, CURRENT ESTIMATED NET COSTS INCLUDE $2.2 BILLION IN CREDIT SUBSIDY, PLUS ADMINISTRATIVE EXPENSES 14 (Apr. 2015), http://www.gao.gov/assets/670/669847.pdf.

[51] *Program Info, Clean Renewable Energy Bonds (CREBS)*, U.S. DEP'T OF ENERGY, http://energy.gov/savings/clean-renewable-energy-bonds-crebs.

bonds (initially providing funds in 2008 and allocating more in 2009); bond issuers like local governments and electric cooperatives apply to Congress for these funds, and bondholders receive "federal tax credits in lieu of a portion of the traditional bond interest, resulting in a lower effective interest rate for the borrower."[52] This is important because unlike other bonds, CREBS "are treated as taxable income for the bondholder."[53]

Tax Credits for Biofuels (including VEETC): Until 2012, the bulk of the renewable energy tax incentives went to promote biofuels.[54] Although some have expired, others remain in place. These past and existing tax credits include: (1) the biodiesel tax credit, which gives a $1.00 per gallon tax credit to producers of pure biodiesel and renewable diesel that meet certain standards, and which have expired several times but in 2015 were extended through December 2016; (2) the "second generation" (i.e., cellulosic) biofuels tax credit, which grants producers a $1.01 per gallon tax credit and, despite earlier expirations, now runs until January 1, 2016;[55] and (3) the Volumetric Ethanol Excise Tax Credit (VEETC) which, until January 2012 when it expired, "granted fuel blenders a tax credit of $0.45 for every gallon of pure ethanol they blended with gasoline,"[56] and on its own resulted in approximately $6 billion in lost revenues in 2010.[57]

2015 Legislation Extending Renewable Energy Tax Credits: As noted above, the PTC and ITC for wind and solar energy had expired or were set to expire by or in 2016. However, the Omnibus Consolidated Appropriations Act,[58] passed by Congress at the end of 2015 and signed into law by President Obama, created a compromise between fossil fuel and renewable energy proponents. The Act lifted the ban on crude oil exports[59] while extending the two major renewable energy credits.[60] With regard to wind energy, the PTC under Section 45 of the Tax Code had expired at the end of 2014. The new law extended the PTC for wind energy facilities for two

---

[52] *Id.*

[53] *Id; see also* Alexandra B. Klass, *Property Rights on the New Frontier: Climate Change, Natural Resource Development, and Renewable Energy*, 38 ECOL. L.Q. 63, 75 (2011).

[54] SHERLOCK & STUPAK, *supra* note 5, at 10, n. 31.

[55] *See* SHERLOCK & STUPAK, *supra* note 5, at 7; U.S. EIA, DIRECT FEDERAL FINANCIAL INTERVENTIONS AND SUBSIDIES IN ENERGY IN FISCAL YEAR 2013, 15–16 & Tbl. 3 (Mar. 2015).

[56] Klass, *supra* note 18, at 33–34.

[57] *Id.*

[58] Pub. L. No. 114–113 (Dec. 18, 2015).

[59] *Id.* at Division O—Other Matters, § 1, tit. 1.

[60] *Id.* at Division Q—Protecting Americans from Tax Hikes Act of 2015, § 1, tit. 1, subtitle C, pt. 3, § 187.

years to include eligible projects beginning construction in 2015 and 2016. After 2016, projects can recover a reduced percentage of the PTC—80 percent for 2017 project starts, 60 percent for 2018 projects starts, and 40 percent for 2019 project starts.[61] For wind projects opting for the ITC over the PTC, the same time extensions and phase-out percentages apply. Other technologies eligible for the PTC, such as geothermal, municipal waste, landfill gas, and biomass projects, received a two-year extension through the end of 2017.[62]

For solar and other renewable energy projects eligible for the ITC under Section 48 of the Tax Code, the 30 percent credit was set to decrease to 10 percent by the end of 2016. The 2015 legislation allowed projects commencing construction prior to 2020 to recover the full 30 percent credit. Projects beginning construction in 2020 are eligible for 26 percent of the ITC, projects beginning construction in 2021 are eligible for 22 percent of the ITC, and projects beginning construction after 2021 are eligible for 10 percent of the ITC, as is any project begun prior to 2021 but not placed in service before 2024.[63] Finally, the legislation extends the existing tax credits for second-generation biofuels production and for biodiesel and renewable diesel until January 1, 2017.[64]

State, Local Government, and Utility Subsidies and Incentives: Many states, local governments, and utilities provide a wide variety of tax incentives and other financial benefits for renewable energy and energy efficiency projects. These include, for example, local and state tax rebates for installing renewable energy systems.[65] As discussed further in section 2 of this Chapter, through the Property Assessed Clean Energy Program these governments also sometimes cover the up-front costs of installing rooftop solar photovoltaic systems, which a homeowner or business owner then pays back over time. For more information on state and local programs, see Database of State Incentives for Renewables and Efficiency (DSIRE), www.dsireusa.org and U.S. Dept. of Energy, Tax Credits, Rebates, and Savings, www.energy.gov/savings.

---

[61] *See* Felix Mormann, *Fading Into the Sunset: Solar and Wind Energy Get Five More Years of Tax Credits with a Phase-Down*, 47 ABA TRENDS 5 (2016); *see, e.g.*, John R. Kirkwood, et al., *Breakdown of Energy and Renewable Energy Provisions in Newly Signed Tax Extenders and Government Funding Legislation*, FAEGRE BAKER DANIELS, Dec. 28, 2015, http://www.faegrebd.com/breakdown-of-energy-and-renewable-energy-provisions-in-newly-signed.

[62] Kirkwood et al., *supra* note 61.

[63] *Id.*; Mormann, *supra* note 61.

[64] Kirkwood et al., *supra* note 61.

[65] *See, e.g., Tax Credits, Rebates, and Savings*, ENERGY.GOV., http://energy.gov/savings/search?f[0]=im_field_rebate_state%3A859946; Klass, *supra* note 53, at 75 (describing more than 20 states that had renewable energy tax credits in 2011).

III. Energy Mandates

Congress and states can encourage certain types of energy development and use not only through use of taxes and subsidies but through mandates that require the generation, sale, or use of certain types of energy resources. This section first discusses mandates and minimum payments for renewable electricity resources such as RPSs, FiTs, and net metering programs. It then moves to the transportation sector and details the federal Renewable Fuel Standard, which mandates the blending of billions of gallons of biofuels into gasoline and diesel fuels sold in the United States.

A. Mandates and Minimum Payments for Renewable Electricity

Renewable Portfolio Standards (RPSs): More than 25 states now have RPSs, which are regulatory mandates that require utilities and other electricity providers to "increase production of energy from renewable sources such as wind, solar, biomass and other alternatives to fossil and nuclear electric generation."[66] Most RPSs require covered electricity providers to generate or purchase an increasing percentage of retail electric energy sales each year (often 15 percent to 20 percent by 2020 or 2030), from renewable energy sources. There is significant variation among the states over the percentage required (California is 50 percent by 2030 while other states are as low as 5–10 percent), which electricity providers must comply, and which renewable resources count toward meeting the RPS. Virtually all states count wind energy, very few count nuclear or existing hydropower energy, and the definition of qualifying biomass differs from state to state. Many states exempt municipal utilities and rural electric cooperatives from compliance. Some states have separate mandates for required percentages of sales from solar energy.[67] RPSs also create compliance structures, which determine how compliance will be assessed and reported, whether there are penalties for failure to comply, and whether the state's public utility commission or legislature has authority to waive or reduce the RPS if electricity prices exceed a particular rate.[68]

Some states have created incentives or requirements for electricity providers to build or procure renewable generation within the state by placing a minimum percentage on in-state renewable

---

[66] *State & Local Governments*, NAT'L RENEWABLE ENERGY LAB., http://www.nrel.gov/tech_deployment/state_local_governments/basics_portfolio_standards.html.

[67] *See* DSIRE, Renewable Portfolio Standards (RPSs) With Solar or Distributed Generation Provisions (summary map) (Aug. 2015).

[68] For detailed information on each state's RPS, see DSIRE, Renewable Portfolio Standards; DSIRE, Renewable Portfolio Standard Policies (summary map) (June 2016).

generation or creating a "multiplier" such as 1.1% or 1.2% for MWs of renewable electricity from in-state sources. Questions arise, however, as to whether such incentives or requirements violate the dormant Commerce Clause of the U.S. Constitution by facially discriminating against out of state renewable energy.[69]

Renewable Energy Credits (RECs): RECs are a key component of the RPSs discussed above because they help entities meet the RPS by creating a "tradable certificate" that can be used in a market to purchase the "environmental benefit" of renewable energy" including (if permitted), from out-of-state sources.[70] RECs generally represent 1 MWh of electricity generated from renewable electricity resources. In many jurisdictions, RPS-covered electricity providers can meet their percentage requirements in full or in part through the purchase of RECs.[71] So if it is more expensive for one utility to generate its own renewable energy resources, it can purchase RECs from another power producer that can generate renewable energy at lower cost. RECs tend to be traded in three different markets—state specific markets, regional markets, and to a lesser extent national markets. The two primary reasons for purchasing RECs are for RPS compliance ("compliance markets") and for green marketing purposes ("voluntary markets").[72] REC prices tend to be much higher in compliance markets ($50–$60/MWh in 2012) than in voluntary markets ($1–$10/MWh in 2012).[73]

In general, when renewable energy is sold from a generator to a utility or other electricity provider, the Power Purchase Agreement (PPA) provides whether the REC remains with the generator and can later sold be separately (known as an "unbundled" REC) or is sold with the electricity to the electricity provider (known as a "bundled" REC).[74] Most wind PPAs in compliance markets provide that the

---

[69] *See, e.g.,* Illinois Commerce Comm'n v. FERC, 721 F.3d 764 (7th Cir. 2013) ("ICC II") (questioning whether Michigan RPS, which prohibits compliance from out of state power, violates the dormant Commerce Clause). For scholarship assessing RPS dormant Commerce Clause issues, *see, e.g.,* Daniel K. Lee & Timothy P. Duane, *Putting the Dormant Commerce Clause Back to Sleep: Adapting the Doctrine to Support State Renewable Portfolio Standards,* 43 ENVTL. L. 295 (2013); Steven Ferrey, *Threading the Constitutional Needle with Care: The Commerce Clause Threat to the New Infrastructure of Renewable Power,* 7 Tex. J. Oil Gas & Energy L. 59 (2011–2012).

[70] Alexandra B. Klass & Elizabeth J. Wilson, *Interstate Transmission Challenges for Renewable Energy: A Federalism Mismatch,* 65 VAND. L. REV. 1801, 1812 (2012).

[71] Alexandra B. Klass, *Takings and Transmission,* 91 N.C. L. REV. 1079, 1119 (2013).

[72] Karlynn Cory, Nat'l Renewable Energy Lab., *Can Renewable Energy Certificates (RECs) Help Your Project Cross the Finish Line?,* Aug. 15, 2011, https://financere.nrel.gov/finance/content/can-renewable-energy-certificates-recs-help-your-project-cross-finish-line.

[73] *Id.*

[74] CAL. ENVTL. PROT. AGENCY, AIR RESOURCES BD., PROPOSED REGULATION FOR A CALIFORNIA RENEWABLE ELECTRICITY STANDARD, STAFF REPORT: INITIAL

REC is sold with the electricity, thus ensuring that the entity that purchases the "green" electricity is the entity that receives the credit. An unbundled REC, in contrast, allows one entity to purchase green electricity and a second entity to purchase the REC and take credit for that green power, which is sometimes considered to involve disingenuous double counting or double claiming.[75] For example, say that a wind generator produces a unit of "green" electricity and sells that electricity to a utility, which uses the electricity purchase to help meet a state RPS. Then, a second entity that is not subject to an RPS mandate—a restaurant, for example—purchases the REC from that same unit of electricity. The restaurant advertises that it is powered by "green" electricity. This would be a somewhat disingenuous claim, since in reality only one unit of electricity had been generated, and this same unit would: (1) count toward the RPS requirement, and (2) count toward the restaurant's voluntary commitment to green electricity. Those who are concerned about double counting would prefer that if the restaurant in this wants to genuinely claim that it is "green," the restaurant should purchase green electricity and retire the REC along with that electricity.

Beyond the issue of counting the sale of electricity as a separate transaction from the sale of a REC (the "unbundling" issue), there is a concern that one REC might be improperly used more than once and thus double counted. States want to avoid having two different utilities count the same REC toward their RPS obligation, for example, and states do this by using REC tracking systems. Once the REC is "used" for compliance or marketing purposes, it is retired and cannot be used again. Regional Tracking Systems are used to track the ownership of RECs, similar to an online bank account. The U.S. Environmental Protection Agency (EPA) describes these systems as follows:

> A tracking system issues a uniquely numbered certificate for each MWh of electricity generated by a renewable generation facility registered in the system, tracks the ownership of certificates as they are traded, and retires the certificates once they are used or claims are made based on their attributes or characteristics. Because each MWh has a unique identification number and can only be in one

---

STATEMENT OF REASONS ES–9 (June 2010), http://www.arb.ca.gov/regact/2010/res 2010/res10isor.pdf.

[75] UNIV. OF PENN. CLIMATE ACTION PLAN 136, http://rs.acupcc.org/site_media/ uploads/cap/82-cap.pdf ("If the physical electricity and the associated RECs are sold to separate buyers, the electricity is no longer considered 'renewable' or 'green.' ").

owner's account at any time, this reduces ownership disputes and the potential for double counting.[76]

Feed-in Tariffs (FiTs): FiTs provides a secure contract for renewable power at a set price designed to ensure that renewable energy developers are able to build generation and sell power from this generation and recover the costs of these projects. FiTs also help ensure that investors are able to make returns on their project investments. To accomplish these goals, FiTs require utilities and other electricity providers to connect renewable generation facilities (often rooftop PV solar) to the electric grid and pay a premium price for the renewable power generated under a long-term contract— typically a contract that lasts for 15 or 20 years. Many FiTs base the rates on the cost of the renewable resources. These characteristics of FiTs help spur investment in new renewable resources because of the certainty associated with the investment as a result of the grid connection and premium price guarantees. FiTs are widespread in Europe, particularly in Germany and Spain,[77] and are used far less frequently in the United States, where RPSs are the dominant policy tool to encourage new renewable energy generation and use.

As of 2015, several states, including California, Hawaii, Vermont, and Washington, have FiTs or similar programs; a few other states have pilot programs or voluntary FiTs that apply to particular utilities; and several cities, including Gainesville, Florida and cities in California, use FiTs.[78] One difficulty with implementing FiTs in the United States is reconciling the premium payment provisions of FiTs with the requirements of the Federal Power Act and PURPA. As explained in Chapter 6, the Federal Power Act gives FERC exclusive jurisdiction to approve wholesale power rates, and PURPA in effect sets the maximum premium price (the "avoided cost") that can be charged for PURPA Qualifying Facility (QF) power sales. "Avoided cost" is the maximum cost to the utility of purchasing power from a generating plant other than the QF or generating the power itself. As noted in Chapter 6, PURPA does not apply in states that are part of regional electricity markets like RTOs, but FERC still has jurisdiction over wholesale sales of electricity in those states and must approve tariff rates. Thus, there are concerns that in states

---

[76] *REC Tracking*, U.S. ENVTL. PROT. AGENCY, https://www.epa.gov/greenpower/renewable-energy-tracking-systems.

[77] TOBY COUTURE & KARLYNN CORY, E3 ANALYTICS & NAT'L RENEWABLE ENERGY LAB., STATE CLEAN ENERGY POLICIES ANALYSIS (SCEPA) PROJECT: AN ANALYSIS OF RENEWABLE ENERGY FEED-IN TARIFFS IN THE UNITED STATES 6 (June 2009), http://www.nrel.gov/docs/fy09osti/45551.pdf.

[78] *Id.* at 7; Database of State Renewables and Efficiency, http://programs.dsireusa.org/system/program/maps (use filter option of "Feed-in Tariff"); *Feed-in Tariffs and Similar Programs*, U.S. ENERGY INFO. ADMIN., June 4, 2013, https://www.eia.gov/electricity/policies/provider_programs.cfm.

where PURPA applies, the PURPA avoided cost rates for QFs (which are set by each state) establish a maximum price for renewable facilities that are also QFs. In states where PURPA does not apply, FERC must approve any FiT.

Nevertheless, in 2010, FERC held that California's program to use a FiT-like mechanism to promote combined heat and power (CHP) generation was not contrary to the Federal Power Act or PURPA so long as the CHP facility also met the requirements for a QF and so long as the rate the California PUC established did not exceed the "avoided cost" of the purchasing utility as required by PURPA.[79] In a later decision, FERC held that the California PUC could "tier" avoided cost rates to differentiate among different QF technologies such as CHP, PV solar, etc., and could also take into account the effect on price of any state RPS requirements.[80]

Net Metering: Net metering is another state program to encourage the installation of renewable energy, particularly PV solar and small wind facilities. Unlike FiTs, which are still rare in the United States, over 40 states have net metering programs.[81] Net metering is a way of crediting entities with renewable energy resources or other types of generation for the excess electricity that they generate and send back to the grid—electricity that is then consumed by other electricity users. For example, a business or home with solar panels on its rooftop often generates more electricity than the business or home uses during very sunny hours, and the electricity meter "spins backward." This excess electricity flows back through the electric distribution wires connected to the business or home, and thus back into the broader grid, to be used by other customers. Rather than paying the business owner or homeowner directly for this excess electricity, the utility provides a credit on the customer's electricity bill, thus reducing the bill.[82] Many states' net metering programs limit the size of generators eligible for net metering (ranging from 10kW to several MWs) or limit the number of generators that can participate in the program (such as 5 percent of the utility's peak demand).[83] Some states, like Hawaii and Nevada, have recently eliminated net metering programs.

---

[79] California Public Utilities Comm'n, 132 FERC ¶ 61,047 (2010).

[80] California Public Utilities Comm'n, 133 FERC ¶ 61,059, at P 26–27 (2010), *reh'g denied*, 134 FERC ¶ 61,044 (2011).

[81] DSIRE, Net Metering (summary map) (Feb. 2016).

[82] *See, e.g.*, DSIRE, Customer Credits for Monthly Net Excess Generation (NEG) Under Net Metering (summary map) (Jan. 2016).

[83] *See* DSIRE, Net Metering Programs, http://programs.dsireusa.org/system/program?type=37& (describing different states' policies); *see also* J. HEETER ET AL., NAT'L RENEWABLE ENERGY LAB., STATUS OF NET METERING: ASSESSING THE POTENTIAL TO REACH PROGRAM CAPS (Sept. 2014), http://www.nrel.gov/docs/fy14osti/61858.pdf.

In practice, using solar PV systems as an example, net metering acts as a way of paying solar customers (typically through credits to their electricity bill, rather than through direct payments) for the excess electricity that they provide to the utility.[84] Unlike FiTs, net metering never results in affirmative payments to generators. However, in some net metering regimes, customers who have net metered a surplus of excess electricity back to the utility at the end of a billing year (because the solar panels "produced more energy than [the customer] consumed during [the] 12-month relevant period") can receive payment for "net surplus energy."[85]

As distributed energy, particularly rooftop PV solar, has increased significantly across the country, many utilities have requested that state PUCs impose new, fixed charges on retail customers with solar PV systems. Utilities argue that by receiving billing credit through net metering programs, customers that generate their own electricity are not paying a sufficient share of the fixed costs of the grid they still use (electric transmission lines, poles, meters, etc.) because those costs are bundled together with energy charges in the billing process. Net metering customers use less electricity and therefore have smaller monthly electric bills; thus, some argue that these customers are failing to adequately pay for the fixed grid costs that they impose. In response, customers who self-generate along with environmental and renewable energy groups argue that no study has yet quantified these additional costs caused by net metering customers and that the distributed generation fed into the grid reduces peak demand and lowers costs for the utilities and their customers. There have been high profile debates on this issue nationwide, with state PUCs imposing significant fixed charges on solar customers in Arizona, Nevada, Wisconsin, and other states at the request of utilities.[86] At least one state, Minnesota, has legislatively created a "value of solar" tariff which is a rate design policy utilities can adopt that creates a solar-specific tariff by which solar customers are paid by the utility that more specifically accounts

---

[84] *See Net Metering*, SOLAR ENERGY INDUSTRIES ASS'N, http://www.seia.org/policy/distributed-solar/net-metering.

[85] *Understanding Your Energy Bill for Net Energy Metering Customers*, S. CAL. EDISON, https://www.sce.com/wps/wcm/connect/b94f53d4-81f5-4e7f-ad0d-1c60c2a1fa64/NEM_FactSheet.pdf?MOD=AJPERES.

[86] *See* Jeffrey Tomich, *Battles Over Fixed Charges Proliferate Across Midwest in Wake of Wis. Cases*, ENERGYWIRE, June 15, 2015; Richard Martin, *Battles Over Net Metering Cloud the Future of Rooftop Solar*, TECH. REVIEW, Jan. 5, 2016. *But see* Gavin Bade, *The Future of Rate Design: Why the Utility Industry May Shift Away From Fixed Charges*, UTILITY DIVE, Nov. 19, 2015, http://www.utilitydive.com/news/the-future-of-rate-design-why-the-utility-industry-may-shift-away-from-fix/409504/ (discussing trends toward a collaborative approach for new rate design that involves a low fixed cost charge, time of use pricing to encourage shift of use to off-peak hours, and inclining block rates to encourage reduced consumption).

for the benefits and costs of solar.[87] Utilities in other states are also considering such tariffs even without legislation.[88]

Energy Efficiency Resource Standards (EERS): An EERS establishes "specific, long-term targets for energy savings that utilities or non-utility program administrators must meet through customer energy efficiency programs."[89] States adopt EERSs through legislation or regulation, and they can apply to electric utilities, natural gas utilities, or both. An EERS is similar to an RPS in that an RPS requires electric utilities to generate a certain percentage of electricity from renewable sources, while an EERS "requires utilities to achieve a percentage reduction in energy sales from energy efficiency measures."[90] Over 25 states have EERS requirements or goals.[91] According to the American Council for an Energy-Efficient Economy, Massachusetts, Rhode Island, and Vermont have the most ambitious EERSs, "which require almost 2.5% savings annually."[92]

B. Mandates for Biofuels

Beyond the tax support for biofuels discussed above, as introduced in Chapter 4, the United States has long buttressed the biofuels industry through the Renewable Fuel Standard (RFS), which requires that a certain amount of renewable fuels be blended with gasoline. Each year, a set amount of these fuels must be included in the gasoline supply, with the amounts to be blended increasing annually through 2022. Pursuant to the Energy Independence and Security Act of 2007 (EISA), a certain volume of the renewable fuel blended with gasoline each year must be from non-corn ethanol, such as cellulosic ethanol made from plant material. However, the statutes that mandate certain quantities of biofuels to be used in the U.S. fuel supply allow the EPA to change these quantities if it appears that regulated entities cannot comply, and the EPA routinely lowers the quantity of non-ethanol fuel required. Specifically, EPA has permission from Congress to

---

[87] *See, e.g., Value of Solar Tariffs*, NAT'L RENEWABLE ENERGY LAB., http://www.nrel.gov/tech_deployment/state_local_governments/basics_value-of-solar_tariffs.html ("Factors that affect VOS rate may include: Utility variable costs (fuel and purchased power); Utility fixed costs (generation capacity, transmission, and distribution); Distribution system and transmission line losses; Ancillary services (to maintain grid reliability); Environmental impacts (carbon and criteria pollutant emissions).").

[88] *See* Kristi E. Swartz, *Debate Over Solar's Value Simmers On in Southeast*, ENERGYWIRE, Jan. 19, 2016.

[89] 79 Fed. Reg. 34,830, 34,872, n.176 (June 18, 2014).

[90] *Energy Efficiency Resource Standards (EERS)*, AM. COUNCIL FOR AN ENERGY EFFICIENT ECON. (ACEEE), http://aceee.org/topics/energy-efficiency-resource-standard-eers.

[91] DSIRE, Energy Efficiency Resource Standards (and Goals) (summary map) (March 2015).

[92] *See* AM. COUNCIL FOR AN ENERGY EFFICIENT ECON. (ACEEE), *supra* note 90.

temporarily waive portions of the biofuels mandate on its own authority or in response to a petition based on market and other circumstances.[93] For instance, EPA may reduce the total amount of biofuels required to be blended for a particular year if there is "inadequate domestic supply" to meet the mandate or if "implementation of the requirement would severely harm the economy or environment of a State, a region, or the United States."[94] Because there was no commercial production of cellulosic ethanol when EISA was enacted, EPA has separate waiver authority to lower the statutory total amounts of cellulosic ethanol when the projected volume for a given year is less than what the statute requires.[95] Although EISA requires that EPA set the required biofuel levels for each year by November of the prior year, EPA has often been unable to comply with that timeline, leading to market and investment uncertainties.[96]

Many members of Congress, the oil industry, vehicle manufacturers, the food and restaurant industry, and some environmental groups have called for the repeal or significant modification of the RFS. First, the RFS is seen as a subsidy for the farming sector at a time where domestic oil is now plentiful as a result of hydraulic fracturing and directional drilling technologies. Second, the RFS is blamed for increasing food prices in the United States and around the world because so much corn is diverted from food production to fuel production. Third, vehicle and engine manufacturers oppose the RFS because of concerns associated with the "blend wall." This term refers to risk of corrosion damage to older vehicle engines if the mix of ethanol in gasoline exceeds its current level of 10 percent.[97] EISA's growing volumetric requirement each year for renewable fuels means that in future years fuel blenders will need to far exceed the blend wall to meet the RFS because U.S. demand for transportation fuel has been flat or decreasing in recent years and is expected to remain that way as a result of the increasing fuel efficiency of new motor vehicles. Fourth, policy analysts and

---

[93] BRENT D. YACOBUCCI, CONG. RESEARCH SERV., WAIVER AUTHORITY UNDER THE RENEWABLE FUEL STANDARD (RFS) 3 (Sept. 25, 2012), https://www.fas.org/sgp/crs/misc/RS22870.pdf.

[94] 42 U.S.C. § 7545(o)(7)(A)(i); YACOBUCCI, supra note 93, at 3–4.

[95] 42 U.S.C. § 7545(o)(7)(D); KELSI BRACMORT, CONG. RESEARCH SERV., THE RENEWABLE FUEL STANDARD (RFS): WAIVER AUTHORITY AND MODIFICATION OF VOLUMES 5–6 (June 29, 2015).

[96] BRACMORT, WAIVER AUTHORITY, supra note 95, at 7 ("Waiver authority can impact RFS implementation and market confidence, as well as contribute to RFS uncertainty. . . . Many aspects of the RFS could be viewed as unsteady . . . partly because Administration decisions—including the use of RFS waiver authority—have not been made in a timely manner.").

[97] Grocery Mfrs. Ass'n v. EPA, 693 F.3d 169 (D.C. Cir. 2012) (discussing concerns by engine manufacturers regarding engine failures in older vehicles and equipment if ethanol content of gasoline exceeds 10 percent); YACOBUCCI, supra note 93, at 28–31.

critics contend that the costs associated with producing the advanced biofuels and cellulosic ethanol amounts required by the RFS will significantly increase fuel prices because, unlike corn ethanol, these biofuels are not cheap or easy to produce.[98] Last, the use of life cycle analysis in evaluating transportation fuels has caused many scientists and policy makers to conclude that ethanol, particularly corn-based ethanol, creates more GHG emissions and other air emissions per gallon than gasoline.[99] Of course, these criticisms are vehemently denied by the biofuels industry, corn states, and others, who maintain that corn ethanol, along with other forms of biofuels, should remain an integral part of the nation's transportation fuels and are environmentally superior to petroleum. The heated debates over the benefits and harms of corn ethanol as well as advanced biofuels will undoubtedly continue in both the science and policy realms. The environmental effects of biofuels are discussed in more detail in Chapter 4.

In addition to calls for repeal or modification of the RFS, the oil industry and other RFS opponents have filed numerous petitions with EPA since the enactment of EISA requesting it to exercise its waiver authority to reduce the mandated amounts of biofuels for each year. EPA has used its waiver authority to significantly reduce the required amount of cellulosic ethanol each year since 2010 based on the lack of sufficient commercial production of such fuel. In 2015, for the first time, it used its waiver authority to also lower the total amount of renewable fuel mandated for 2014 (retroactively) as well as for 2015 and 2016, citing "blend wall" concerns and the biofuel industry's inability to produce sufficient volumes of advanced biofuels.[100] Oil companies and other interested parties have sued EPA numerous times for its refusal to exercise its waiver authority more frequently, with mixed results in the courts.[101]

---

[98] *See, e.g.,* Terry Dinan, Congressional Budget Office, Testimony: The Renewable Fuel Standard: Issues for 2015 and Beyond at 2–3 (Nov. 3, 2015) (in one scenario estimating costs, noting that "meeting the requirements [for volume] . . . would have significant effects on prices of transportation fuels").

[99] *Id.* at 20–21.

[100] *See* U.S. Envtl. Prot. Agency, Renewable Fuel Standard Program: Standards for 2014, 2015, and 2016 and Biomass-Based Diesel Volume for 2017, 80 Fed. Reg. 77,420 (Dec. 14, 2015) (to be codified at 40 C.F.R. pt. 80); *Final Renewable Fuel Standards for 2014, 2015 and 2016, and the Biomass-Based Diesel Volume for 2017,* U.S. ENVTL. PROT. AGENCY, RENEWABLE FUEL STANDARD PROGRAM, https://www.epa.gov/renewable-fuel-standard-program/final-renewable-fuel-standards-2014-2015-and-2016-and-biomass-based#rule-summary; Jim W. Rubin, *Fueling Controversy: EPA Seeks the Right Balance But its Final RFS Volumes Are Likely To Be Challenged,* DORSEY & WHITNEY PUBL'NS, Jan. 8, 2016, https://www.dorsey.com/newsresources/publications/client-alerts/2016/01/epa-seeks-the-right-balance.

[101] *See, e.g.,* Monroe Energy, LLC v. EPA, 750 F.3d 909 (D.C. Cir. 2014); American Petroleum Institute v. EPA, 706 F.3d 474 (D.C. Cir. 2013); Mfrs. Ass'n v. EPA, 693

IV. Energy Financing

Many of the instruments described above—tax credits, FiTs, and RPSs, for example—provide much of the certainty that investors need before deciding to embark upon an expensive renewable energy development project. However, companies moving forward with these types of projects also rely on other mechanisms to ensure that they will in fact make money from a project. They also need creative means of ensuring that they can benefit from policy incentives like tax credits. For example, a small wind or solar company with low tax liability often cannot take full advantage of tax credits, therefore requiring mechanisms for temporarily partnering with companies that can benefit from these credits, or tools that encourage investment even when tax credits are unavailable. This section briefly discusses some of the contractual instruments and other private law approaches used by renewable energy developers in order to take advantage of some of the policy incentives introduced above and to augment these incentives. Many of these approaches include public law elements because certain financing instruments have been created and enabled by federal and state statutes and judicial decisions. For example, one means of financing rooftop solar panels—in which local governments pay for the up-front costs of the panels and recover these costs by adding a charge to a homeowner's annual property taxes—is an approach enabled by state legislation.[102] This section discusses approaches to financing utility-scale (large, centralized) renewable generation projects as well as distributed (on-site generation) projects.

A. Financing for Utility-Scale Renewable Energy Infrastructure

One of the greatest challenges faced by developers of large-scale renewable generation projects is the inability to fully benefit from tax credits like the PTC for wind energy described above. This is particularly true for relatively small, independent companies that are only in the business of building renewable generation, as compared to large utilities that build many types of power plants and have more revenues, costs, and tax liability.[103] One common way in

---

F.3d 169 (D.C. Cir. 2012); Nat'l Petrochemical & Refiners Ass'n v. EPA, 630 F.3d 145 (D.C. Cir. 2010).

[102] *See, e.g.*, Vernon's Texas Statutes and Codes Annotated, Local Government Code § 376.004(a) (allowing local legislative bodies to "determine that it is convenient, advantageous, and in the public interest to designate an area of the municipality within which authorized municipal officials and property owners may enter into contracts to assess properties to finance the installation of distributed generation renewable energy sources or energy efficiency improvements that are permanently fixed to real property").

[103] *See* Felix Mormann, *Beyond Tax Credits: Smarter Tax Policy for a Cleaner, More Democratic Energy Future*, 31 YALE J. ON REG. 303, 325 (2014) (noting that

which smaller companies avoid this hurdle is to work with a tax-equity investor, meaning an investor that obtains a "substantial portion of the return" on investment through tax credits or other tax benefits[104] rather than, say, dividends or similar cash distributions.

In one example of tax-equity investment, the renewable energy developer partners with a larger company for a limited time through a mechanism called a "partnership-flip."[105] The renewable energy developer—the smaller company—creates a partnership in the form of a limited liability corporation (LLC) with the larger company.[106] Through this LLC arrangement, the developer maintains control over all day-to-day construction and similar decisions, and the larger company with relatively high tax liability—the tax equity investor—merely serves as the investor and is not as involved in decisions about the project.[107] The large investor, which has provided a substantial chunk of money to the renewable energy developer, secures this investment using the tax equity (the investor's right to receive the tax benefits from the project) "rather than a direct mortgage or security interest in the wind plant assets."[108] The large investor is able to take advantage of the tax credit because tax income can "be allocated in a manner that is different from the ownership percentages."[109] Further, to allow the investor eventual exit from the partnership once the tax credits expire (often after ten years, in the case of a wind project), the LLC is formed in a manner such that over time, the percentage ownership interest of the tax equity investor declines—typically when the investor has achieved a certain target of tax savings.[110] Eventually, the renewable energy developer owns the entire project. By this time both parties have benefited from the transaction because the large company that invested in the project as a partner provided needed money for the project—thus aiding the smaller renewable energy company—and the larger company

---

"renewable energy developers and their projects tend to lack the quintessential requirement to benefit from tax credits—a high enough tax bill to offset with these credits"); David Burton & Joe Sebik, *Taking Some of the Mystery Out of Alternative Energy Flip Structures* at 40, EQUIP. LEASING & FIN. MAGAZINE (May/June 2013), https://www.akingump.com/images/content/2/3/v4/23301/FinancialWatchMayJune20 13.pdf ("Due to the high cost of wind farms and the large tax benefits created, developers often are unable to utilize the tax benefits efficiently and, accordingly, need to implement a flip partnership.").

[104] Greg Jenner & David T. Quinby, Energy Project Development and Finance Seminar: Tax Issues & Opportunities at 17, University of Minnesota Law School, Oct. 29, 2015 (on file with authors).

[105] Mormann, *supra* note 103, at 331 (describing this term and its meaning).

[106] Burton & Sebik, *supra* note 103, at 40.

[107] *Id.*

[108] Stoel Rives, The Law of Wind: A Guide to Business and Legal Issues, ch. 7, p. 5 (2014) (on file with authors).

[109] Burton & Sebik, *supra* note 103, at 40.

[110] *Id.*

received tax savings.¹¹¹ This type of "flip" is allowed due to an IRS Revenue Procedure issued in 2007 that creates a "safe harbor" for using this type of partnership for the production tax credit.¹¹²

Another advantage of the partnership flip described above—as well as other types of partnerships designed to support certain large-scale renewable energy projects—is that it allows the members of the partnerships to be subject to lower taxation in the form of "pass-through" taxation. Specifically, in the partnership flip described above, the "the LLC does not owe tax, rather the tax benefits and tax liabilities 'flow through' to the partners."¹¹³ But Professor Felix Mormann also notes the limits of tax equity investor instruments such as the partnership flip, observing that any financing mechanisms that rely on Congressionally approved tax credits are unpredictable and that only a few companies are large enough to serve as the tax equity investor and to fully benefit from the tax credit, among other challenges.¹¹⁴

There are financing alternatives to tax equity investment that avoid these challenges, and Congress has proposed to make these mechanisms more readily available to renewable energy developers—although they are not readily available now. Pass-through taxation similar to that seen in the partnership flip example above can be advantageous beyond the tax equity context because it can encourage a larger and more diverse set of investors to provide needed capital for a renewable energy project, even without the carrot of a tax credit. For example, through a Master Limited Partnership (MLP), which allows for pass-through taxation, one or two investors serve as the general investors that control the partnership and typically have a two percent ownership interest in it, while a larger number (potentially thousands) of other investors called "unitholders" or limited partners invest money and receive quarterly cash distributions but do not manage the partnership.¹¹⁵ As Professor Mormann explains, these distributions are like dividends, and all of the cash from the project is distributed to the unitholders with the exception of money needed to finance the business.¹¹⁶ MLPs are currently not available to renewable energy developers. Although members of Congress have introduced the "MLP Parity Act" several times—an act that would extend the MLP

---

¹¹¹ *See* Mormann, *supra* note 103, at 325 ("In essence, the tax equity investor's capital contribution buys her the rights to the project's tax benefits—and helps the developer finance the project's high up-front capital expenditures.").

¹¹² Burton & Sebik, *supra* note 103, at 40; Internal Revenue Bulletin: 2007–45, Rev. Proc. 2007–65, Nov. 5, 2007.

¹¹³ Burton & Sebik, *supra* note 103, at 40.

¹¹⁴ Mormann, *supra* note 103, at 326–31.

¹¹⁵ Mormann, *supra* note 103, at 341.

¹¹⁶ *Id.*

mechanism to renewable energy—to date those efforts have not succeeded.[117]

A similar financing mechanism provides tax benefits that can also attract large numbers of smaller investors, but it, too, is currently of limited use to renewable energy developers. This mechanism—the real estate investment trust (REIT)—is similar to the MLP in that it allows a company to attract small investors. Specifically, it allows "small investors to invest in a professionally managed pool of real property in a tax-efficient manner."[118] A REIT does not allow automatic pass-through of taxation directly to investors. However, if the trust is created in a certain way—with a large percentage of its assets being in real estate and mortgages—and if the trust distributes a larger percentage of its income to shareholders, many of the dividends distributed to shareholders are taxed based on shareholders' individual income (not the trust's income), thus providing a tax advantage.[119] Renewable energy developers have used REITs,[120] but the extent to which developers can rely on this mechanism remains uncertain, particularly because it is still not fully clear which assets of the renewable project count as real property that can be included in the trust.[121] It is clear, however, that assets like the land underlying the wind farm and the towers that support wind turbines count as real property that can be included within a REIT.[122]

B. Financing for Distributed Renewable Energy Infrastructure

In addition to large, centralized renewable power plants, distributed renewable energy installed on rooftops, in backyards, and in parking lots is increasingly common, and creative financing mechanisms for this type of energy are expanding. Distributed solar, in particular, has grown astronomically. For example, the nation's largest installer and owner of rooftop solar—SolarCity—experienced approximately 60 percent revenue growth between 2014 and 2015,

---

[117] *MLP Parity Act Reintroduced*, BAKER BOTTS (June 2015), http://www.bakerbotts.com/ideas/publications/2015/06/mlp-update.

[118] Scott A. Bank & Kelly Myers Kogan, REITS & Renewables: An Emerging Combination, Chadbourne & Park LLP, http://www.chadbourne.com/files/upload/Kogan-REITS-Solar-Presentation.pdf (PowerPoint presentation), at 4.

[119] Mormann, *supra* note 103, at 342–43.

[120] Kelly Kogan, *The Status of REITs and Renewables: An Update from the Field*, SOLARINDUSTRYMAG.COM, http://solarindustrymag.com/online/issues/SI1309/FEAT_04_The_Status_Of_REITs_And_Renewables_An_Update_From_The_Field.html (noting that "REITs are finding ways to include renewable energy assets in their portfolios").

[121] Bank & Kogan, *supra* note 118, at 16, 19.

[122] *Id.* at 19.

although its costs also grew during this period.[123] Generation from small-scale solar photovoltaic (PV) energy now makes up approximately one third of all solar energy generation in the United States.[124]

Owners of homes and businesses have increasingly flocked to distributed solar—typically in the form of rooftop PV panels—for environmental reasons, and, in some cases, to gain access to a cheaper supply of electricity than is available from the large public utility that serves the home or business. The cost of rooftop solar has recently plummeted dramatically, due in large part to booming production of PV panels in China.[125] But even with declining costs, the up-front bill for purchasing and installing solar panels can be quite high. Homeowners and businesses have a variety of options for addressing these costs.

Owners who do not wish to self-finance the up-front cost of a rooftop solar system can obtain a special bank loan for the solar panels.[126] Some loans are equity-based,[127] whereas others are cash-type loans that do not require appraisals of the value of a home or business.[128] Alternatively, in the approximately thirty-one states property owners can benefit from a program called "Property Assessed Clean Energy"[129] (PACE). For PACE to be an option, a state must first allow the use of this instrument by local governments. Specifically, the state must permit the local government to enter into contracts with homeowners or business owners, in which the local government agrees to pay the up-front costs of solar panel installation, and the owner agrees to repay the local government over time through a fee assessed as part of the owner's annual property

---

[123] Diane Cardwell & Julie Creswell, *SolarCity and Other Rooftop Providers Face a Cloudier Future*, N.Y. TIMES, Feb. 10, 2016, http://www.nytimes.com/2016/02/11/business/energy-environment/rooftop-solar-providers-face-a-cloudier-future.html?_r=0.

[124] *EIA Electricity Data Now Include Estimated Small-scale Solar PV Capacity and Generation*, ENERGY INFO. ADMIN., (Dec. 2, 2015), http://www.eia.gov/todayinenergy/detail.cfm?id=23972.

[125] GALEN BARBOSE & NAÏM DARGHOUTH, LAWRENCE BERKELEY NAT'L LAB., TRACKING THE SUN VIII: THE INSTALLED PRICE OF RESIDENTIAL AND NON-RESIDENTIAL PHOTOVOLTAIC SYSTEMS IN THE UNITED STATES 1–2 (Aug. 2015) (noting that between 2013 and 2014 installed PV system costs declined by 9 percent and that similar declines were expected to continue, due largely to the "steep drop in global prices for PV modules").

[126] *See* Nichola Groom, *Loans Challenge Big Money's Leasing Model for U.S. Rooftop Solar*, REUTERS, Sept. 24, 2013, http://www.reuters.com/article/us-solar-loans-idUSBRE98N04C20130924.

[127] *See, e.g.*, *Solar Panel Home Equity Loans*, ALTS. FED. CREDIT UNION, http://www.alternatives.org/solarpanelloans.html.

[128] *See, e.g.*, *Admirals Bank's Solar Loans*, ADMIRALS BANK, https://www.admiralsbank.com/renewable-energy-lending/loan-programs/solar-step-down.

[129] *See* List of all PACE enabling statutes by state, PACENation, http://www.pacenation.us/resources/pace-enabling-legislation/.

taxes, or through a separate fee assessed at different time intervals. For example, in New Jersey, the state allows local governments to do the following:

> [T]he governing body of a municipality may undertake the financing of the purchase and installation of renewable energy systems ... by property owners as a local improvement and may provide by ordinance for a "clean energy special assessment" to be imposed on a property within the municipality, if the owner of the property requests the assessment in order to install such systems or improvements. Each improvement on an individual property shall constitute a separate local improvement and shall be assessed separately to the property owner benefitted thereby. The clean energy special assessment shall be payable in quarterly installments.[130]

This is particularly advantageous to the property owner because if the owner moves and sells the property, he or she will not have paid for the full value of the solar panels, only to potentially lose some of that value in the sale. Rather, the new owner of the property will pay for the remaining costs of the system through annual property taxes.

Another common means of financing rooftop solar is third-party ownership in the states where this model is allowed. Through third-party ownership a company owns the solar panels and simply uses a home or business as a location on which to house the solar panels. The company enters into a PPA with the owner of the property, in which the company commits to sell the electricity from the solar panels to the property owner for a set amount of money per kilowatt-hour over a long-term period, such as twenty years. Sometimes the PPA provides that the price per kilowatt-hour will escalate by a fixed amount during this time period. The property owner receives the benefit of a predictable, fixed price for electricity—often lower than the rate charged for electricity by the large local utility. Further, the third-party owner does all installation and maintenance of the panels over time and covers all installation and maintenance costs in addition to the up-front costs of the panels. This allows the property owner to avoid these responsibilities and costs. The third-party owner, in turn, gets the benefit of a guaranteed customer who will purchase electricity over a long period of time.

Third-party ownership is the most popular financing model for rooftop solar.[131] However, as of 2016, at least nine states effectively

---

[130] N.J. STATS. § 40:56–1.4.
[131] GEORGINA ARREOLA ET AL., CTR. FOR SUSTAINABLE ENERGY & LAWRENCE BERKELEY NAT'L LAB., SURVEY OF BUYERS, SELLERS AND REALTORS INVOLVED IN SAN

disallowed third-party ownership by including third-party solar companies within the definition of a "public utility"—which would require these utilities to go through a complex process of receiving approval to operate within another utility's service territory and to have their rates and infrastructural decisions approved—or by allowing net metering only for systems owned by the property owner.[132] In many other states, the law is unclear whether third-party solar companies would be considered "public utilities," creating disincentives for solar companies to offer such arrangements in these states because of concerns over violating laws protecting existing utilities' monopoly territories.[133]

Another form of third-party ownership financing is the solar lease, in which a company leases solar panels to the property owner. Similar to the third-party ownership model, the up-front costs of installation are covered by the company, and property owner—using the lease—can ensure that the solar company conducts ongoing maintenance of the panels. But the owner does not have the option of purchasing electricity from the company at a fixed rate over time; rather, the owner must make periodic payments simply to maintain ownership status of the panels. Additionally, the lease rate sometimes escalates over time, depending on the terms of the lease.[134]

The range of public support mechanisms for renewable energy as well as private financing tools is expanding for both utility-scale and distributed systems, but barriers remain. For utility-scale energy, these barriers include, inter alia, a lack of clarity regarding how renewable projects can quality as a REIT and the failure to allow MLPs for these projects. For distributed energy, many states have not yet definitively indicated whether third-party ownership is allowed, and several essentially prohibit it.[135] By lifting these types of barriers, the federal government and state governments could meaningfully incentivize more investment in renewable energy projects.

DIEGO THIRD-PARTY OWNED SOLAR HOME TRANSACTIONS—A QUALITATIVE ASSESSMENT 1, https://emp.lbl.gov/sites/all/files/lbnl-1003917.pdf (noting that "[b]y 2013, third-party owned (or 'TPO') systems dominated the [national] market, representing over 70% [of] residential installations").

[132] Samuel Farkas, *Third-Party PPAs: Unleashing America's Solar Potential*, 28 J. LAND USE & ENVTL. L. 91, 101–06 (2012); DSIRE, 3rd Party Solar PV Power Purchase Agreement (July 2016) (summary map).

[133] See DSIRE, *supra* note 132.

[134] NATE HAUSMAN, CLEAN ENERGY STATES ALL., A HOMEOWNER'S GUIDE TO SOLAR FINANCING: LEASES, LOANS, AND PPAS 5–6 (2015), http://www.cesa.org/assets/2015-Files/Homeowners-Guide-to-Solar-Financing.pdf.

[135] See DSIRE, *supra* note 132.

Chapter 6

ENERGY MARKETS

I. Introduction

Courts and legislatures at both the federal and state levels have historically treated the markets that make up the energy industry as unique markets. Specifically, most companies that provided fuels or electricity were labeled as "public utilities." Private investors owned many of these utilities (indeed, they were and are commonly called "investor-owned utilities," or IOUs), but governments treated them as "public" utilities because of the essential good that they provided to the public. Courts and legislatures bestowed unique authority on these utilities over time, giving them a leg up in markets. Many of them received eminent domain authority so that they could build needed infrastructure even over private landowners' objections.[1] And states and the federal government also chose to treat them as "natural monopolies," which are businesses for which necessary infrastructural investments are very costly, thus causing high barriers to entry. Natural monopolies also are, in some cases, more efficiently operated by one or just a few entities. For instance, rather than having numerous electric transmission line companies competing for customers and constructing redundant lines, it may be superior to have a more limited amount of infrastructure to supply electricity transmission needs.

By treating energy industries as natural monopolies, states and the federal government created both benefits and burdens for these industries. On the benefits side, a business that is a natural monopoly and is regulated as a natural monopoly receives government permission to be the sole provider of a particular energy good within a particular region, or "service territory." In other words, the government mandates that this entity operate as a monopoly, prohibiting incursions by competitors into its service territory. This provides the industry with a guaranteed set of customers. However, on the burdens side, the government places service obligations on the natural monopoly. The company must serve all customers who demand service within the territory, and it typically may not cut off customers for a failure to pay without first providing notice and an additional opportunity for the customers to pay. Further, the rate that the utility may charge its customers is a regulated rate, which is capped by the government regulator. The rate is typically a "cost

---

[1] *See* Alexandra B. Klass, *Takings and Transmission*, 91 N.C. L. REV. 1079, 1105 (2013).

of service" rate. This means that the government regulator determines how much money the utility needs to incur debt and attract investors, build infrastructure, and operate its facilities, and sets a rate designed to ensure that if the utility continues operating relatively efficiently and sells similar amounts of electricity or fuel to the amounts sold in the past, it will recover these costs through the rates that it charges. Finally, because the government allows public utilities to recover their costs by charging retail rates, the government wants to ensure that major costs incurred by the utility are worthwhile costs, and that the utility does not build more infrastructure than is necessary to serve customers. Therefore, governments that treat utilities as natural monopolies must approve utilities' infrastructural investments—such as decisions to build a new power plant or pipeline—before the utilities make these investments. This infrastructural approval (approval to build a power plant, for example) is called a "certificate of need" or a "certificate of public convenience and necessity." Once a utility has obtained a certificate of need approving the construction of infrastructure, it also goes through a "ratemaking proceeding" to receive approval for recovering the costs of attracting investment in, building, and operating this infrastructure.

Public utilities have not always been regulated as natural monopolies. For example, in the first quarter of the 20th century, the electric industry tended to involve many small power plants that served a small portion of a city's population—Chicago alone had power plants run by more than forty different firms.[2] But as the industry began to consolidate, with large companies buying up numerous electric generation plants, various lobbyists engaged in a nationwide effort to convince states that electric utilities should be regulated as natural monopolies.[3] Similarly, at the federal level, as the federal government gained control over wholesale sales (i.e., from one utility to another for later resale to a retail customer) of electricity,[4] interstate transmission of electricity, wholesale sales of natural gas,[5] and rates charged by interstate pipelines,[6] the federal

---

[2] Robert L. Bradley, Jr., *The Origins and Development of Electric Power Regulation*, in THE END OF A NATURAL MONOPOLY: DEREGULATION & COMPETITION IN THE ELECTRIC POWER INDUSTRY 43, 73 n.4 (Peter Z. Grossman and Daniel H. Cole, eds., 2003).

[3] *Id.* at 46–48.

[4] 16 U.S.C. § 824(a) (1935). (The statute remains largely the same today, and FERC continues to exercise this authority.)

[5] Phillips Petroleum Co. v. Wisconsin, 347 U.S. 672 (1954) (interpreting the Natural Gas Act to require what was then the Federal Power Commission (now FERC) to regulate wholesale natural gas sales in which producers sold gas to buyers who eventually transported the gas interstate). Congress eventually relieved FERC of this responsibility in the Wellhead Decontrol Act of 1989.

[6] Alexandra B. Klass & Danielle Meinhardt, *Transporting Oil and Gas: U.S. Infrastructure Challenges*, 100 IOWA L. REV. 947, 960 (2015) (describing the Hepburn

government more heavily regulated these transactions and treated many aspects of the industry as natural monopolies. For example, the federal government gave natural gas pipeline companies certain territories in which they were allowed to operate, determined when and how much infrastructure they could build, and capped the rate that these pipeline companies could charge for the natural gas that they owned and transported through their pipelines.[7] Similarly, the federal government capped the rate that one electric utility could charge another for a wholesale sale of electricity.[8]

Over time, the federal government and some states have treated fewer and fewer portions of the energy industry as natural monopolies. This change is sometimes described as "deregulation" but is more accurately described as "restructuring." Governments still regulate the energy industry, but the level and scope of regulation have generally declined in certain states.[9] Some states, however, have retained traditional natural monopoly models and still fully regulate certain energy industries. This Chapter uses electricity as a primary example of energy markets at the federal and state levels to explain how the federal and state governments regulate these markets, and how these markets have changed over time as a result of restructuring. It then briefly compares electricity markets to natural gas and oil markets.

II. Federal Regulation of Energy Markets

In order to understand energy markets and government regulation of them, one must first have a basic knowledge of the different physical components of the system for extracting or generating energy, moving it long distances, and distributing the energy to customers. Each of these physical components involves different types of contracts, prices, market practices, and regulation.

The market for electricity consists of three specific parts: the generation of electricity and wholesale sales of this electricity to other utilities, transmission of electricity from generation plants to

---

Act of 1906, which provided for federal regulation of interstate oil pipelines under the Interstate Commerce Act, thus creating federal control over interstate oil pipeline rates and terms of service, but not construction and siting); 15 U.S.C. § 717(b) (West 2016) (providing for federal regulation of interstate natural gas pipelines).

[7] *See* JAMES H. MCGREW, FERC: FEDERAL ENERGY REGULATORY COMMISSION, BASIC PRACTICE SERIES 55–57 (2003) (noting FERC's role in granting certificates of need to pipelines to approve their construction and setting the rates that pipelines and natural gas providers could charge based on cost-of-service ratemaking).

[8] *See* California ex rel. Lockyer v. Fed. Energy Regulatory Comm'n, 383 F.3d 1006, 1012 (9th Cir. 2004) (noting FERC's "historical policy of basing rates upon the cost of providing service plus a fair return on invested capital" in the wholesale electricity context).

[9] *Status of Electricity Restructuring by State*, U.S. ENERGY INFO. ADMIN (2010), http://www.eia.gov/electricity/policies/restructuring/restructure_elect.html.

utilities that have purchased the electricity, and distribution of electricity from utilities to retail customers. Companies that generate electricity at plants fueled by coal, natural gas, wind, solar, hydropower, nuclear, or other energy resources enter into contracts called power purchase agreements (PPAs)—which are typically long-term (fifteen-or twenty-year contracts)—with other utilities who commit to purchasing the power at a set rate over this period. Some generators, in addition to or instead of entering into PPAs, sell their electricity through auctions for wholesale electricity that are available in some parts of the country where regional transmission operators (RTOs) or independent system operators (ISOs) control a large, regional, interconnected grid of electric transmission lines and determine which generators may send a certain quantity of electricity through the grid at a particular time. Generators that have committed to sell electricity under a contract or through an auction contract with a transmission line owner to use space within the transmission line, and the rate that the generator pays to the transmission line owner is a federally regulated rate.[10] Finally, utilities that generate their own electricity (referred to as "vertically integrated utilities") or purchase electricity for resale sell this electricity to retail customers such as businesses, residents, and industries; these utilities charge a retail rate that is regulated by the state. In some cases states have chosen to restructure retail electricity markets and to allow the retail rate to be set by market forces rather than a state regulatory agency.

Natural gas markets are similar to electricity markets. Companies that extract natural gas enter into contracts with purchasers—often purchasers who are far from the point of natural gas extraction. As with electricity, these contracts for the wholesale sale of natural gas are often long-term contracts. Purchasers of natural gas are typically large power plants, which use natural gas to generate electricity, and "local distribution companies" (LDCs), such as cities, which purchase natural gas and then distribute it at retail to customers for heating, other residential uses, and industrial uses. In some states, customers that were once retail customers of LDCs are allowed to directly contract with gas producers for gas; large industrial users are the most common entities to take advantage of this option. Producers of natural gas who have entered into contracts with wholesale purchasers then contract with a pipeline company to use space in the pipeline to transport natural gas. Alternatively, a "gas marketer"—an entity in the business of connecting wholesale gas buyers and gas sellers—also sometimes arranges for the transport of natural gas, contracts with the pipeline

---

[10] 16 U.S.C. § 824(a) (1935).

for transportation.[11] Regardless of the entity that contracts with the pipeline company to use space in the pipeline to transport natural gas, the price that the entity pays to the pipeline is federally regulated. Finally, the price that local distribution companies charge retail customers when they sell gas to homes, businesses, and industrial users is regulated by the state. As with electricity, in some cases states have stopped regulating the retail rate of natural gas due to a restructuring of the regulatory system.

Oil markets are similar to natural gas markets, although the entities that purchase oil are much more varied. A substantial amount of oil produced in the United States goes to refineries and is then sold overseas for export.[12] Furthermore, although many oil producers contract with pipelines to transport their product, a growing percentage of the oil produced in the United States—particularly in North Dakota—is transported by rail because of inadequate pipeline capacity in certain regions of the country.

This section explores the federal regulation of the energy markets just introduced. It focuses primarily on electricity but also briefly discusses federal regulation of natural gas and oil.

A. Federal Regulation of Electricity Markets

When electricity first became a commonly-offered service, there was no federal regulation of the industry at all. States and local governments regulated power plants, transmission lines, and distribution lines. In many cases, utilities lobbied local governments to acquire a franchise allowing them to provide electricity to customers within the local government's jurisdiction.[13] However, states began to more heavily regulate utilities in order to protect customers (and because utilities lobbied for regulated natural monopoly status).[14] States granted utilities exclusive service territories within regions of the state and regulated the rates that they could charge. But as states more heavily regulated utilities—especially utilities that engaged in interstate transactions—this began to create interstate conflicts.

Interstate conflicts over utility regulation came to a head in the *Public Utilities Commission v. Attleboro Steam and Electric Co.* case,

---

[11] *Gas Marketing*, NATURALGAS.ORG, http://naturalgas.org/naturalgas/marketing/.

[12] The United States banned *crude* oil exports (as opposed to exports of refined oil), with limited exceptions, from 1975 through 2015. Pub. L. no. 94–163 (1975); 15 C.F.R. § 754.2(b) (West 2016). Congress only recently lifted the ban. Consolidated Appropriations Act, 2016, H.R. 2029. For further discussion of the history of the ban and the recent lifting of the ban see Chapter 7.

[13] Bradley, *supra* note 2, at 73, n.4.

[14] *Id.* at 46–48.

decided by the U.S. Supreme Court in 1927.[15] In this case, Narragansett Electric Lighting Company was a vertically-integrated investor-owned utility that operated in Rhode Island, meaning that the company owned and operated power plants, transmission lines that it used to move its own electricity through certain parts of the state, and distribution lines that the company used to distribute electricity to individual homes and businesses.[16] The State of Rhode Island fully regulated Naragansett Electric Lighting Company. It determined how many power plants, transmission lines, and distribution lines the company could build and how much the company could charge its retail customers through rates designed to recover Naragansett's costs of generating electricity, transporting electricity, and distributing electricity to customers.[17] Thus, when Naragansett proposed to sell excess electricity that it generated—electricity not needed by retail customers in Rhode Island—to an electric utility in Massachusetts, Rhode Island had to approve this sale.[18] Rhode Island did approve this sale, establishing the rate that Naragansett could charge for its wholesale sale of electricity to the Massachusetts utility—Attleboro Steam and Electric Company.[19]

Naragansett and Attleboro entered into a long-term contract for this wholesale electricity transfer, and Attleboro then decommissioned its own power plant in Massachusetts, thinking that it could simply rely on Naragansett for continued supply of electricity to sell to Attleboro's retail customers in Massachusetts.[20] However, Naragansett's costs of generating electricity rose, and Naragansett therefore tried to raise the rates that it charged Attleboro for electricity under the contract. Rhode Island approved this higher wholesale rate,[21] and Attleboro appealed, arguing that Rhode Island's order allowing a Rhode Island utility to charge a higher rate for a sale of electricity to a Massachusetts utility was a direct burden on interstate commerce.[22] The Rhode Island Supreme Court and the U.S. Supreme Court agreed. The U.S. Supreme Court found that the wholesale sale of electricity across state lines was "essentially national in character" and could not be regulated by the states.[23] But at the time, there was no federal regulation of electricity, and this therefore created a gap called the *Attleboro* gap—neither states nor

---

[15] 273 U.S. 83 (1927).

[16] *Id.* at 84.

[17] *Cf. id.* (noting that Naragansett had to apply to the Rhode Island Public Utilities Commission to receive approval of its proposed rate schedule change).

[18] *Id.*

[19] *Id.* at 84–85.

[20] *Id.* at 85.

[21] *Id.* at 85–86.

[22] *Id.* at 86.

[23] *Id.* at 90.

the federal government regulated wholesale sales of electricity in interstate commerce.

Following the creation of the *Attleboro* gap, Congress enacted the Federal Power Act of 1935, which was not a new act but rather an expansion of the Federal Water Power Act of 1920—an Act that had primarily focused on federal regulation of hydroelectric dams by an agency called the Federal Power Commission (FPC). The Federal Power Act gave the FPC new authority, providing for federal regulation of the "transmission of electric energy in interstate commerce and the sale of such energy at wholesale in interstate commerce."[24] Thus, beginning in 1935 the federal government regulated all electricity sales like the sale of electricity from Naragansett to Attleboro both because such a sale is a "wholesale sale" of electricity (from one utility to another) and also because such a transaction is for the "transmission of electric energy in interstate commerce."

In that same year, Congress also further regulated power plants that were built to generate electricity for both retail and wholesale sales. Congress was concerned that very large holding companies were buying up and consolidating too many power plants, creating monopolies that were dangerous despite only being allowed to charge regulated rates for the electricity sold. Congress thus enacted the Public Utility Holding Company Act of 1935 (PUHCA), which required utilities of a certain size—utilities that counted as utility "holding companies" because they owned so much generation and other assets—to register with the Securities and Exchange Commission,[25] obtain approval before issuing securities,[26] and obtain approval before buying or selling other public utilities,[27] among other requirements.

A key aspect of FERC's role under the Federal Power Act was to set the rates that utilities could charge for selling wholesale power to other utilities, and to separately approve the rates that utility owners of transmission lines could charge for the use of those lines. As introduced above, FERC historically set these rates based on a cost-of-service ratemaking model. Under this model, FERC determines how much it costs the utility to provide the particular service for which it is charging the rate, such as producing and selling electricity (a wholesale sale) or transporting electricity for another entity. These costs fall under three broad categories: (1) capital costs associated

---

[24] 16 U.S.C. § 824(a) (1935).
[25] 15 U.S.C. § 79d (repealed 2005).
[26] 15 U.S.C. § 79f (repealed 2005).
[27] 15 U.S.C. § 79d (repealed 2005).

with the service, called the "rate base,"[28] which, in the case of wholesale power primarily involves the costs of the power plant, and in the case of transmission lines involves the costs of the physical wires; (2) the rate of return that the utility must pay to investors in order to incentivize them to invest in capital, defined as " 'the amount of money earned by a public utility, over and above operating costs, expressed as a percentage of rate base' ";[29] and (3) the operational costs associated with providing the service, such as paying employees to run and maintain a power plant or transmission lines, and other expenses such as taxes. These three components are often expressed as the following ratemaking formula: cost of service of a utility = "Rate Base x Overall Rate of Return = Return [paid to investors] + Operation & Maintenance Expenses" in addition to other expenses such as taxes.[30] FERC calculates these costs—the value of the rate base, operational expenses, etc.—based on real costs that the utility has incurred in a prior year called a "test year." Based on this prior year, FERC estimates what the utility's costs are likely to be in future years, taking into account anomalies in the test year that might not apply in future years or additional expenses that might arise in future years that were not reflected in the test year.[31]

After calculating the total cost of service incurred by a utility, FERC determines how to ensure that the utility has an opportunity to recover those costs by charging a per-unit rate, such as a rate per megawatt-hour of electricity sold or transported. The rate set by FERC does not guarantee that the utility will recover all of its costs; if it ends up operating less efficiently than it has in the past, for example, it cannot automatically raise its capped rates and might end up losing money. But the rate gives the utility an opportunity to recover all of its costs and provide a reasonable rate of return to investors if it operates under the constraints under which FERC assumed it would operate when FERC set the rate. In a simplified hypothetical example of how FERC gives the utility an opportunity to recover rates (say, for a wholesale sale of electricity), assume that a utility's total rate base value in the test year was $2 million, the rate of return that the utility needed to pay to investors in order to attract investment in capital was 10%, and the utility's total operating and administrative costs, taxes, etc. were $1 million. The total cost of service would be (.10 x $2 million) + $1 million, or $1,200,000. FERC would then estimate the utility's total number of

---

[28] DAVID J. MUCHOW & WILLIAM A. MOGEL, ENERGY LAW & TRANSACTIONS 4–17 (1990).

[29] Id. at 4–19 (quoting CHARLES F. PHILLIPS, THE REGULATION OF PUBLIC UTILITIES 332 (2d ed. 1985)).

[30] FED. ENERGY REGULATORY COMM'N, COST-OF-SERVICE RATES MANUAL 6 (1999).

[31] MUCHOW & MOGEL, supra note 28, at 52–24.

wholesale customers and their total electricity purchases and divide the cost of service among these purchases. Say the utility had 100 wholesale customers, each of whom annually purchased 100 megawatt-hours of electricity in a year. If FERC set the rate at $120 per megawatt-hour of electricity sold, the utility would recover the following: $120/megawatt-hour x 100 wholesale customers x 100 megawatt-hours/customer = $1,200,000. In reality, however, FERC does not divide up the cost of service equally among customers. Instead, it divides up costs among classes of customers before setting the rates. Customers that cause the utility to incur more costs—e.g., customers that require a fixed amount of electricity to be delivered whenever electricity is demanded, and thus require the utility to build "peaker" plants to supply electricity during periods of peak demand—might pay a rate that reflects a larger percentage of the costs of the peaker plants.

When FERC sets rates based on a utility's cost of service, the Federal Power Act requires that these rates be "just and reasonable."[32] Once FERC sets a rate for the amount of money that a utility may charge, this is a "filed rate," meaning that the utility has proposed a rate to the Commission, and the Commission has officially approved the rate, typically with revisions from the utility's proposal. Once the Commission has approved the rate, the reasonableness of the rate may only be challenged through FERC, not through the courts—with only limited exceptions. This is called the "filed rate doctrine."[33]

Although all wholesale rates must be approved by FERC, utilities may enter into private contracts and set their own rates, and a different standard of review from the "just and reasonable" standard applies to these rates. For example, Utility A may enter into a PPA with Utility B, in which the two utilities agree on a rate. The utilities in the past simply had to submit this rate to FERC for approval, and FERC could only disapprove the rate if it were unjust and unreasonable. Similarly, FERC could only modify or abrogate the rate if a party challenged the rate and FERC found that it was contrary to the public interest.[34] In other words, although FERC retained (and still retains) the power to reject or modify a rate that parties have agreed upon in a private contract, this power is

---

[32] 16 U.S.C. § 824d (West 2016).

[33] Texas Commercial Energy v. TXU Energy, 413 F.3d 503, 507 (5th Cir. 2005) ("The filed rate doctrine bars judicial recourse against a regulated entity based upon allegations that the entity's 'filed rate' is too high, unfair or unlawful.").

[34] David G. Tewksbury & Stephanie S. Lim, *Applying the Mobile-Sierra Doctrine to Market-Based Rate Contracts*, 26 ENERGY L.J. 437, 456 (2005) ("In cases involving cost-based rate contracts, the courts have repeatedly and consistently held that the Mobile-Sierra doctrine allows the Commission to modify contracts only if required by the public interest" (internal quotations omitted)).

somewhat limited; the purpose of this doctrine is to protect private contract integrity while also ensuring that entities that might have monopolistic power do not charge unfair rates.[35] This doctrine still applies today and is called the *Mobile-Sierra* doctrine, which was established by two Supreme Court cases—one addressing electricity rates and the other addressing natural gas rates.[36]

For many years after the enactment of the Federal Power Act and PUHCA, utilities largely continued to be vertically integrated and regulated under the cost-of-service ratemaking model. One utility company, which we will call "Utility A" for simplicity, built its own power plants and transmission lines, used its own lines to move electricity among its power plants and to sell wholesale electricity to other companies, and built and operated its own distribution lines to transmit electricity to retail customers. The federal government regulated the rate that Utility A could charge when it sold electricity to other utilities (Utilities B and C, for example). And if Utility A ever allowed another utility to use its transmission lines, the federal government regulated the rate that Utility A could charge the other utility for the use of the lines but had more limited authority over the utility's other actions and transactions, which created market problems.

One example of the limited authority that the FPC (and later FERC) initially had over transmission lines was the inability of the agency to order wheeling. When one utility allows another utility to use its transmission lines to transport electricity to a third utility this is called "wheeling." Assume that Utility B has a contract to sell electricity to Utility C, but Utility B does not own and operate its own transmission lines and therefore has no way to transport the

---

[35] *Id.* at 467. The limited ability of FERC to modify rates set by private contracts was affirmed by the U.S. Supreme Court in *Morgan Stanley Capital Group v. Public Utility District No, 1 of Snohomish Cty.*, where the Court reiterated that FERC must presume contract-based rates to be just and reasonable unless the "contract seriously harms the public interest." 554 U.S. 527, 530 (2008). The court also noted that this presumption applies even when FERC did not initially have the opportunity to review the contract because the contract involved a "going" market rate that FERC "passively" allowed, such as a price for electricity. Regardless of when FERC reviews the contract, the standard is whether the rate harms the public interest. *Id.* at 546–47. However, the court interpreted the public interest standard in a manner that was slightly beneficial to those challenging the reasonableness of rates, noting that FERC should consider whether certain types of rates "imposed an excessive burden on consumers 'down the line,'" not just immediately. *Id.* at 552. Further, "if it is clear that one party to a contract engaged in such extensive unlawful market manipulation as to alter the playing field for contract negotiations, the Commission should not presume that the contract is just and reasonable." *Id.* at 554.

[36] *Id.* at 437–43 (noting that the doctrine "bars the Commission from reforming or abrogating a fixed-rate contract absent a showing that contract reformation or abrogation is required to protect the public interest" and describing United Gas Pipe Line Co. v. Mobile Gas Serv. Corp., 350 U.S. 332 (1956) and Fed. Power Comm'n v. Sierra Pac. Power Co., 350 U.S. 348 (1956)).

electricity to Utility C. Utility B could contract with Utility A to "wheel" electricity through Utility A's lines to Utility C. And the rate that Utility A could charge Utility B for use of the lines would be federally regulated. However, for many years utilities that owned and operated transmission lines could simply refuse wheeling requests, and they often did. After all, why would Utility A want to allow Utility B to use Utility A's transmission lines to sell electricity to Utility C? This would increase competition; Utility A would prefer to simply sell electricity to Utility C and not give Utility B that opportunity. The lack of access to transmission also stymied the growth of new competition in the generation of electricity for wholesale sales.

This anticompetitive electricity market lasted for many years but began to change with the U.S. Supreme Court's 1973 decision in *Otter Tail Power Co. v. United States*.[37] Otter Tail Power Company was a large utility that sold electricity throughout the upper-Midwest, primarily to utilities that were owned and operated by municipal governments.[38] These municipal governments purchased their electricity from Otter Tail, but they wanted to purchase cheaper electricity from federal hydroelectric power plants owned and operated by the U.S. Bureau of Reclamation.[39] However, Otter Tail refused to wheel this power to the municipalities.[40] It also engaged in other anti-competitive practices in an attempt to defend its position as the primary wholesale power provider to the municipalities.[41] The municipalities sued in U.S. district court, which held that antitrust law applied to these types of transactions.[42] The district court also prohibited Otter Tail from refusing to wheel the Bureau of Reclamation's power.[43] The U.S. Supreme Court upheld this decision.[44]

Although *Otter Tail* made clear that a court could order wheeling under antitrust law, for more than a decade after the decision the Federal Power Commission—renamed as the Federal Energy Regulatory Commission (FERC) soon after *Otter Tail*—was unsure whether the Federal Power Act gave it to the authority to order a utility to wheel another utility's electricity, despite two

---

[37] 410 U.S. 366 (1973).
[38] United States v. Otter Tail Power Co., 331 F. Supp. 54, 57 (D. Minn. 1971).
[39] *Id.* at 57–60.
[40] *Id.*
[41] *Id.*
[42] *Id.* at 65.
[43] United States v. Otter Tail Power Co., 1971 WL 584 at *1 (D. Minn., Nov. 10, 1971).
[44] *Otter Tail*, 410 U.S. 366, 381.

provisions in the Federal Power Act that appeared to do so.[45] One post-*Otter Tail* act created limited wheeling authority, although only for small generators, and only in limited circumstances. The Public Utility Regulatory Policies Act of 1978 (PURPA) allowed FERC to order utilities to wheel electricity from certain small generators that generated efficient electricity or were renewable generations, but FERC first had to find that several relatively strict conditions were met before ordering wheeling for these types of generators.[46] FERC began slowly encouraging utilities to open up their transmission lines, however, by conditioning its approval of utility mergers on these utilities' promising to grant open access to their transmission lines.[47]

FERC's authority to open up transmission and make it more competitive became clearer when Congress enacted the Energy Policy Act of 1992, which directly gave FERC the power to order wheeling.[48] But FERC still had to order utilities to open up their lines on a case-by-case basis, thus still making it difficult for individual generators to obtain access to a utility's lines in order to wheel their power. As also discussed in Chapter 3, the major change occurred in 1996, when FERC issued Order 888.[49] This order required all utilities that owned and operated transmission lines to make these lines "open access," meaning any generator that wanted to use the lines had to have the opportunity to do so, on a first-come, first-served basis, until there was no more room in the lines. Further, if the utility that owned and operated the transmission line transmitted both its own electricity and other utilities' and generators' electricity, it was not allowed to charge itself more favorable rates for the use of its lines. FERC also issued accompanying Order 889 at the same time.[50] This order required utilities that owned and operated transmission lines to publish an "open access transmission tariff" (OATT) that provided transparent information on the rates charged for use of the transmission lines, thus providing competing utilities and generators (not just the transmission-owning utility) full access to information about how much it would cost to transmit electricity.

---

[45] F. Paul Bland, *Problems of Price and Transportation: Two Proposals to Encourage Competition from Alternative Energy Sources*, 10 HARV. ENVTL. L. REV. 345, 401–402 (1986).

[46] *Id.* at 401–02 (describing PURPA's limited wheeling requirements).

[47] Sheila S. Hollis, *The Changing Scene: An Overview of Recent Developments in the Electric Regulatory Arena*, 42B RMMLF–INST 13 (1996).

[48] Energy Policy Act of 1992, § 721 (1992) (amending section 211 of the Federal Power Act to read: "Any electric utility, Federal power marketing agency, or any other person generating electric energy for sale for resale, may apply to the Commission for an order under this subsection requiring a transmitting utility to provide transmission services. . . .").

[49] 75 FERC ¶ 61,080 (Apr. 24, 1996).

[50] 75 FERC ¶ 61,078 (Apr. 24, 1996).

In Order 888 FERC further encouraged the opening up of the electricity transmission monopoly by suggesting (but not requiring) that transmission line owners give up individual control over their lines to a regional organization called an Independent System Operator (ISO). This created a network of more interconnected transmission lines operated by one entity. This, in turn, made it easier for individual electricity generators to enter into contracts for use of the transmission lines to transport their electricity over long distances to a variety of markets. Previously, these generators had to work with each individual transmission line owner to access each line needed to transport the electricity long-distance. With each transmission owner having a different rate (charge) for the use of its line—a rate approved by FERC—this could result in "pancaked" rates, meaning that the generator wishing to transport its electricity long-distance had to contract with different transmission line owners and pay these owners different FERC-approved rates. By having individual line owners give up operational control over their lines to a regional entity, FERC created a system in which generators could work with one entity to obtain transmission service and avoid pancaked rates. FERC encouraged even more regionalization of the grid through FERC Order 2000 in 1999, which encouraged transmission line owners to cede control over their transmission lines to entities called regional transmission organizations (RTOs), which were essentially the same as ISOs but had to meet several additional requirements to be approved by FERC as an official RTO.[51] Approximately two-thirds of U.S. electricity sales now take place within RTOs and ISOs, which are widespread throughout the United States except in the Southeast and Intermountain West.[52] (For discussion of RTOs' and ISOs' role in planning for the siting of transmission lines, see Chapter 3.)

At this point within the technical maze of federal transmission line restructuring, readers might wonder why utilities that own transmission lines would have agreed to give up operational control over the lines to an RTO or ISO. The answer was partially one of regulatory ease. Any utility that owns and operates a transmission line must obtain a tariff (which is essentially a license) from FERC setting forth the terms under which the utility must offer its service and approving the rate that the utility may charge for the use of its transmission line. Obtaining a tariff can be a time-consuming process. But under Order 888 encouraging ISO formation and Order

---

[51] 89 FERC ¶ 61,285 (Dec. 20, 1999).

[52] ISO/RTO COUNCIL, THE VALUE OF INDEPENDENT REGIONAL GRID OPERATORS 7 (2005), http://www.nyiso.com/public/webdocs/media_room/press_releases/2005/isortowhitepaper_final11112005.pdf; *Regional Transmission Organizations (RTO) /Independent System Operators (ISO)*, FED. ENERGY REGULATORY COMM'N, http://www.ferc.gov/industries/electric/indus-act/rto.asp.

2000 encouraging RTO formation, utilities that turned over control over their lines to an entity approved by FERC as an ISO or RTO no longer had to obtain an individual tariff. Rather, the ISO or RTO that took over control of each utility's lines obtained one, large tariff from FERC for all of the transmission lines that it controlled. This system remains in place today.

RTOs and ISOs play an important role in encouraging competition beyond their offering of regionalized, open access transmission. In order to schedule the amount of electricity that will flow through regional lines at a given time and determine exactly where this electricity will come from and when, RTOs and ISOs also run auctions for electricity. They accept bids from "load-serving entities" (LSEs)—companies that must provide electricity to meet demand (load)—for the amount of electricity that these LSEs will need at a particular time. They also accept offers from generators around the RTO or ISO region; generators offer to provide a particular amount of electricity at a particular rate. RTOs or ISOs then accept all of the bids up to the amount of electricity that is needed, taking the cheapest generation first, and the clearing price for electricity is the price of the last marginal unit of electricity that was needed to meet demand in a given auction.[53] Thus, RTOs and ISOs create a competitive market for electricity itself, rather than just opening up the transmission lines through which electricity flows.

While Congress and FERC were encouraging the opening up of transmission lines in order to allow a larger variety of utilities and generators to have access to the infrastructure necessary to get their products to markets, these entities were also directly encouraging competition in generation. Even with more access to transmission lines, many business entities simply were not adequately incentivized to build new power plants that would compete with the big, vertically integrated utilities that had traditionally built power plants. Thus, Congress in 1978 enacted the PURPA, which was designed to directly incentivize the construction of new, competitive generation while also encouraging energy conservation and renewable energy.[54] Specifically, the Act defined certain generators as "qualifying small power production facilities" (in short "qualifying facilities," or QFs), and these QFs received special treatment under the Act.[55] QFs included (and still include) power plants that were "cogenerators," meaning they generated electricity and used the waste heat or steam for another function, such as heating a building,

---

[53] For a discussion of the RTO auction process, see Fed. Energy Regulatory Comm'n v. Elec. Power Supply Ass'n, 136 S. Ct. 760, 768–69 (2016).

[54] 92 Stat. 3117 (1978).

[55] Public Utility Regulatory Policies Act § 201 (1978).

and facilities that had a capacity of generating up to 80 megawatts of electricity from "biomass, waste, renewable resources," or a combination of these resources.[56] Utilities were required to purchase electricity from QFs, and they were required to purchase this electricity at "avoided cost"—the amount of money that it would have cost utilities to generate the power themselves or purchase it from another provider. QFs also could apply to FERC for an order requiring a utility to allow a QF to connect to the utility's transmission lines so that the QF could sell power to the utility,[57] although, as mentioned above, the power of FERC to require wheeling of QF electricity was not clear at the time. PURPA remains in place today, although in areas of the country where QFs have access to competitive wholesale electricity auctions or to an RTO or ISO, utilities are exempt from PURPA requirements.[58]

Another means that Congress used to encourage more competition in generation was to exempt certain generators from PUHCA requirements introduced above. Recall that PUHCA required public utilities of a certain size to register with the SEC and obtain approval before issuing securities or merging with other utilities. In the Energy Policy Act of 1992, Congress exempted certain generators from PUHCA. Specifically, Congress created a category of generators called "exempt wholesale generators" (EWGs, pronounced "ewogs"), which were generators that were wholly independent of large, vertically integrated utilities.[59] In other words, if a company was only in the business of generating electricity, and not in the business of transmitting and distributing electricity, this company was an EWG and was not subject to PUHCA. This exemption encouraged a great deal of independent generation to be built but is no longer relevant because in the Energy Policy Act of 2005 Congress repealed most parts of PUHCA.[60]

Beyond PURPA and EWGs, FERC encouraged more competition in electricity generation by eventually eliminating the requirement that each electric generator selling wholesale electricity obtain approval for the price that it charged for wholesale electricity. Rather, FERC granted broad-based permission to utilities to sell wholesale electricity at whatever rate the market would bear or a

---

[56] *Id.*

[57] *Id.* at § 210.

[58] 16 U.S.C. § 824a–3 (West 2016).

[59] Energy Policy Act of 1992, § 32 (1992).

[60] FED. ENERGY REGULATORY COMM'N, FACT SHEET: ENERGY POLICY ACT OF 2005 at 4 (2006), https://www.ferc.gov/legal/fed-sta/epact-fact-sheet.pdf (explaining that Congress in 2005 repealed PUHCA 1935 and replaced it with a new PUHCA 2005, which, as summarized by FERC, "permits Commission access to books and records of holding companies and their numbers if necessary for determining jurisdictional rates").

purchaser would agree to in a contract, provided the utilities demonstrated that they operated in competitive markets. FERC did this by issuing a "blanket" certificate to generators and utilities that were deemed to not have market power (to not have monopolistic control within an area), and this certificate broadly allowed these utilities or generators to sell wholesale electricity at the market rate.[61] FERC retained the power to review the utilities' sales under "market-based" rates, however, to determine whether gaming of the market was occurring. Because FERC still technically approves these market rates, the market rates are still filed rates, meaning that if parties wish to challenge the rates as being unjust and unreasonable due to gaming of markets, they have to go through FERC.[62] As a result, when competition in wholesale power was ramping up and companies like Enron gamed the system (for example, shutting down certain generating plants to create artificially high wholesale rates), parties had to challenge these market rates through FERC proccedings.[63] Further, the *Mobile-Sierra* doctrine applies to these rates, meaning that FERC may only modify them if necessary to protect the public interest.[64] To further encourage competition in electricity generation, FERC over time developed rules for operating the RTO and ISO competitive wholesale electricity auctions introduced above—markets in which the price of wholesale electricity is determined by various bids rather than set by FERC.[65]

Thus, current federal electricity markets are largely restructured. Utilities that own transmission lines may not act like true monopolies because they must offer open access to these lines. Further, most wholesale sales of electricity (at least those in RTOs and ISOs) are truly competitive—the price of wholesale electricity is set in auction-based markets. And even outside of RTOs and ISOs, prices for wholesale electricity are set by private contracts (power purchase agreements) between generators and utilities. But FERC

---

[61] Tewksbury & Lim, *supra* note 34, at 447 (in 2005, noting: "Over a decade ago, the FERC 'departed from its historical policy of basing rates upon the cost of providing service plus a fair return on invested capital, and began approving market-based [rate] tariffs' of public utilities. Under this market-based rate regime, the FERC grants a wholesale electricity seller blanket authorization to make sales at market-based (i.e., negotiated) rates only if the seller can demonstrate" a lack of market power (quoting California ex rel. Lockyer v. FERC, 383 F.3d 1006, 1012 (9th Cir. 2004)).

[62] *See, e.g.*, Public Utility District No. 1 of Snohomish Cty. v. Dynegy Power Marketing, Inc., 384 F.3d 756, 762 (9th Cir. 2004) ("FERC approved tariffs that governed the California wholesale electricity markets. Therefore, if the prices in those markets were not just and reasonable or if the defendants sold electricity in violation of the filed tariffs, Snohomish's only option is to seek a remedy before FERC.").

[63] *Id.*

[64] Tewksbury & Lim, *supra* note 34, at 456.

[65] For a description of FERC's gradual move toward competitive wholesale electricity rates, *see* Gerald Norlander, *May the FERC Rely on Markets to Set Electric Rates?*, 24 ENERGY L.J. 65 (2003).

still regulates the price that utilities may charge for the use of their transmission lines because transmission lines remain somewhat of a "bottleneck" (i.e., a monopolistic or capacity constrained segment of an industry). Although utilities must allow anyone to use their transmission lines, provided there is space in the lines, FERC cannot require utilities to build new lines. Space in the lines is therefore limited, and if FERC did not regulate the rate that utilities could charge for use of the lines, utilities could act like a monopolistic toll bridge operator, allowing everyone over the bridge but charging a high price. The following section explores how federal regulation of natural gas markets is very similar to that of electricity markets. Natural gas was initially largely unregulated at the federal level, it was then heavily regulated federally, and current federal natural gas markets are largely restructured.

B. Federal Regulation of Natural Gas Markets

Like the electricity industry, the natural gas industry was initially largely vertically integrated. A given natural gas company was typically in the business of drilling for and producing natural gas, owning pipelines through which it transported its own natural gas, and then selling natural gas to customers. Also similar to the electricity industry, natural gas was initially primarily regulated by states, which created an interstate commerce problem. In 1924, in *State of Missouri ex rel. Barrett v. Kansas Natural Gas. Co.*, the U.S. Supreme Court considered the situation in which natural gas companies that produced gas (primarily in Oklahoma), transported it interstate through pipelines, and sold it in Kansas and Missouri were subjected to different regulations in different states.[66] Specifically, these companies attempted to charge higher rates for their natural gas, asking both the Missouri and Kansas regulators to allow them to raise rates from 35 cents to 40 cents per cubic-foot of gas sold in each state. Missouri refused the request, and Kansas granted it, thus creating higher rates for wholesale purchasers of natural gas in Kansas.[67] As with the *Attleboro* case, the Supreme Court held that when it came to producing natural gas, transporting it interstate, and selling it in various states for a particular wholesale price, "the paramount interest is not local but national," and that states could not regulate these transactions.[68] But at the time the federal government did not regulate the interstate transportation of natural gas or wholesale sales of natural gas in interstate commerce, and this left a gap.

---

[66] 265 U.S. 298, 305 (1924).
[67] *Id.*
[68] *Id.* at 309.

As with electricity, Congress filled this gap by establishing federal jurisdiction over wholesale sales of natural gas and the rates charged by interstate pipelines in the Natural Gas Act of 1938.[69] But in contrast to electricity, where there is virtually no federal jurisdiction over the siting of transmission lines (see Chapter 1), Congress also granted to the federal government siting authority and eminent domain authority for interstate natural gas pipelines in the Natural Gas Act.[70] Thus, FERC currently regulates the following activities: (1) the construction and siting of interstate natural gas pipelines—FERC must grant a "certificate of public convenience and necessity" to the pipeline company determining that the pipeline and associated infrastructure such as compressor stations are needed to serve customers or increase competition and approving the route and construction; (2) the rates that these pipeline companies may charge to customers who use the pipeline; and (3) wholesale sales of natural gas (although as with electricity, FERC generally allows these sales to be at a market-based rate yet retains jurisdiction to monitor possible market manipulation by gas marketers).

Also similar to electricity markets, FERC restructured gas markets over time. Just as FERC initially encouraged and then required transmission lines to be "open access" in FERC Order 888—thus making these transmission lines less like a monopoly—FERC followed a similar process for natural gas pipelines. FERC initially encouraged pipeline companies to offer open access transmission service in Order 436 in 1985, which made it easier for the pipeline companies to obtain FERC approval for constructing pipelines and associated infrastructure if they promised to offer the pipeline on an open access, non-discriminatory basis.[71] This meant that pipeline companies wanting this benefit had to promise to allow anyone to use the pipeline on a first-come, first-served basis provided that there was space in the pipeline, and when the companies were transporting their own gas through the pipeline they could not charge themselves more favorable rates than they charged other companies. In Order 636 issued in 1992, FERC required all interstate natural gas pipelines to provide open access, non-discriminatory service.[72] Pipeline companies also had to provide transparent, electronic information about the rates that they charged for the use of their pipeline. Further, companies had to separate out two types of pipeline service that they provided: "firm" service—providing guaranteed space in the pipeline for an uninterrupted supply of gas,

---

[69] 15 U.S.C. §§ 717, *et seq* (West 2016).

[70] *See* Klass & Meinhardt, *supra* note 6, at 994–98 (discussing reasons why Congress transferred siting and eminent domain authority over interstate natural gas pipelines from the states to the federal government).

[71] 1982–85 FERC ¶ 30,665 (1985).

[72] 59 FERC ¶ 61,030 (1992).

a service that is valuable for customers who need a constant flow of gas to a plant or other facility—and "interruptible" service, for which customers typically are able to use the pipeline to transport gas but might be interrupted when there is inadequate space in the pipeline.

Just as FERC has jurisdiction over wholesale sales of electricity (in addition to the transmission of electricity in interstate commerce), FERC under the Natural Gas Act also controls wholesale sales of natural gas.[73] For nearly three decades, FERC struggled to determine how to accurately set the price of wholesale natural gas. After initially attempting to regulate each natural gas producer as a utility and setting individual prices for each producer, for some time it used regional pricing schemes based on estimates of the costs of producing natural gas in certain regions and thus an estimate of how much money producers needed to recover.[74] However, it was difficult for FERC to accurately estimate prices that should be charged for natural gas. Further, natural gas production had largely become a competitive market and was not a natural monopoly—particularly as access to pipelines grew, thus opening up opportunities to transport gas to various markets. Congress therefore changed the regulatory system for wholesale sales of gas. In the 1980s, Congress made clear that FERC should not control the price of natural gas sold from producers to an entity that purchased the gas "at the wellhead" (gas as it came out of the well) and eventually sold it interstate.[75] Wholesale sales of natural gas owned by pipelines remained regulated[76] until FERC "unbundled" pipelines (requiring them to functionally separate the businesses of owning gas and transporting it).[77] Now gas pipelines do not typically own the gas transported through the pipeline.[78] FERC also eventually stopped regulating the price of natural gas sold from a producer to an interstate gas marketer or a purchaser in another state,[79] although it retained jurisdiction to review these sales for market manipulation and still has the authority to correct rates that it deems to be unjust or unreasonable.

One notable market challenge that has arisen now that FERC encourages competitive wholesale markets for both electricity and

---

[73] 15 U.S.C. § 717(b) (West 2016).

[74] MCGREW, *supra* note 7, at 56–57; In re Permian Basin Area Rate Cases, 390 U.S. 747 (1968) (allowing FERC to establish rates by geographic area).

[75] Natural Gas Wellhead Decontrol Act of 1989, Pub. L. no. 101–60.

[76] FED. ENERGY REGULATORY COMM'N, ENERGY PRIMER: A HANDBOOK OF ENERGY MARKET BASICS 17 (2015), http://www.ferc.gov/market-oversight/guide/energy-primer.pdf.

[77] *Id.* at 17, 32; McGrew, *supra* note 7, at 60–61.

[78] FED. ENERGY REGULATORY COMM'N, *supra* note 76, at 32.

[79] *Id.* at 34 (describing the current, highly competitive market and associated market rates).

natural gas is a lack of coordination between these two markets. Wholesale electricity auctions run by RTOs and ISOs are on a different schedule than are transactions for natural gas transportation and sales. (Those in the electricity business sometimes complain that natural gas markets are on "bankers' hours," whereas electricity markets operate constantly.) Thus, when natural gas-fired electricity plants bid into an RTO or ISO auction indicating that they can provide a certain amount of electricity at a certain price, it can be difficult for these generators to properly formulate their bid when they do not know exactly how much gas they will have available to them at a particular price. Further, natural gas plants with inadequate access to gas as a result of market coordination failures can cause reliability problems. Therefore, in 2012 FERC requested comments on what role it should play, if any, in coordinating natural gas and electricity markets.[80] In 2015, in Order 809, FERC took on a coordinating role, requiring certain timelines in the electricity and natural gas markets, including timelines for natural gas transportation through pipelines, to be better aligned.[81]

C. Federal Regulation of Oil Markets

FERC also has regulatory authority over interstate oil pipelines and markets, but this authority is much more limited. As shown in the discussions above, utilities that generate electricity wholesale and own and operate transmission lines are still considered "public utilities," which FERC regulates to various degrees. Although FERC no longer caps the rate that these utilities may charge for the electricity that they sell wholesale, it approves the rate that utilities may charge for providing access to their transmission lines, determining how much it costs the utilities to provide transmission line service and setting a rate designed to allow the utility to recover most of the costs of providing service. Similarly, FERC must approve natural gas pipeline companies' construction of pipelines before they are built, and FERC also approves the rate that these "public utilities" may charge of customers who use the pipelines.[82]

FERC treats wholesale sales of oil and oil pipelines very differently. Although FERC has jurisdiction over oil pipelines under the Interstate Commerce Act, FERC does not regulate oil pipelines

---

[80] FED. ENERGY REGULATORY COMM'N, REQUEST FOR COMMENTS OF COMMISSIONER MOELLER ON COORDINATION BETWEEN THE NATURAL GAS AND ELECTRICITY MARKETS, Feb. 3, 2012, http://www.ferc.gov/industries/electric/indus-act/electric-coord/moellergaselectricletter.pdf.

[81] 151 FERC ¶ 61,049 (Apr. 16, 2015).

[82] FERC approves pipelines not only because of rates but because siting authority was transferred to it in the NGA because of state roadblocks to interstate pipelines. *See supra* note 70 and accompanying text.

as public utilities and therefore does not apply cost of service ratemaking to these pipelines. As a result, FERC need not determine that they are "needed" before they are built,[83] since the pipelines will not use a FERC-approved rate to recover the costs of building and operating the pipelines from customers. Instead, oil pipelines are true "common carriers," meaning that various shippers individually contract with the pipelines for use of the pipeline. However, FERC requires that the pipelines offer equal access/equal service terms to these shippers and, through an index system, lightly regulates the price that the pipelines may charge of these shippers.[84] The index sets a ceiling for the prices that pipelines may charge and is based on the median and mean five-year historic pipeline costs and the producer price index for finished goods.[85] One problem previously experienced by oil pipeline companies was one of uncertainty; the companies would build a pipeline but would sometimes have trouble attracting contracts and thus a guarantee of future financing because of investor uncertainty about rates that would ultimately be set by FERC under the index system. To address this problem, FERC has issued declaratory orders prior to oil pipeline construction approving set rates proposed by the pipeline company, including rates that do not exactly track FERC's indexing system.[86]

One jurisdictional question with respect to oil pipelines has become more important with the recent hydraulic fracturing boom. The boom has not just resulted in the production of more oil and gas from shales and tight sandstones in the United States; fracturing has also produced more natural gas liquids (petroleum substances that are similar to natural gas but exist in liquid form) and related products such as ethane. These products are very valuable in manufacturing processes, and companies that transport these products have recently petitioned FERC for determinations that some of these products are not "oil" and thus that the pipelines carrying the products are not subject to FERC's regulation of oil pipelines. When FERC addresses this type of petition, the central question is whether the product being carried through the pipeline is defined as "oil" under the Interstate Commerce Act. This hinges on whether the primary purpose of the substance is an "energy" purpose—i.e., whether the substance is used to generate energy.[87]

---

[83] Christopher J. Barr, *Growing Pains: FERC's Responses to Challenges to the Development of Oil Pipeline Infrastructure*, 28 ENERGY L.J. 43, 51 (2007).

[84] *Regulating Oil Pipelines*, FED. ENERGY REGULATORY COMM'N, http://www.ferc.gov/industries/oil.asp.

[85] *Oil Pipeline Index*, FED. ENERGY REGULATORY COMM'N, http://www.ferc.gov/industries/oil/gen-info/pipeline-index.asp; Order Establishing Index Level, 153 FERC ¶ 61,312 (Dec. 17, 2015).

[86] Barr, *supra* note 83, at 52–54.

[87] 145 FERC ¶ 61,303 (2013), http://www.ferc.gov/CalendarFiles/201312 31134350-OR13-29-000.pdf at 4 (noting that the "governing test" under the Interstate

FERC already has determined that natural gas liquids count as "oil" under the ICA,[88] but one company recently petitioned FERC for a determination that the substance "purity liquid ethane," similar to a natural gas liquid, was not oil. The company lost, with FERC finding that purity liquid ethane, a type of natural gas liquid, "is a naturally-occurring hydrocarbon product with current energy uses and future undeveloped energy uses."[89]

As noted above, a large amount of oil in the United States is also transported by rail. For example, in February 2015, more than half of the crude oil sent to east coast refineries in 2015 was shipped by rail.[90] And in July 2013, "monthly average carloadings of oil and petroleum products were near 16,000 carloads per week,"[91] with much of this oil moving from North Dakota's Bakken Shale to locations throughout the rest of the United States, but particularly to Gulf Coast oil refineries. Like oil pipelines, rail companies that carry crude oil are "common carriers." Historically, rail companies were required to provide non-discriminatory service to any entities that wished to use rail companies—particularly when rail companies acted like common carriers, meaning "when they undertook to carry goods for all those who applied."[92] But as Christopher Barr explains, rail carriers also may enter into private contracts with companies and may contract away much of their capacity. Thus, rail carriers publish tariffs—a rate that they set and publish for anyone who wants to use the rail line—but also contract with individual shippers.[93] Contracts typically incorporate tariff terms and thus are influenced by the tariffs.[94] Although the tariffs need not be approved by the agency that regulates economic aspects of railways—the Surface Transportation Board—shippers that use railroads may argue before the Board that tariffs are discriminatory.

Commerce Act is "whether the product being transported serves an energy-related, as opposed to a feedstock, function.").

[88] Michelle T. Boudreaux & Amy L. Hoff, *Determining the Jurisdictional Status of NGL Pipeline Transportation Service* at 3, http://cblpipelinelaw.com/news/articles/DeterminingJurisdictionalStatusNGLPipelines_09242013.pdf.

[89] 145 FERC ¶ 61,303, *supra* note 87, at 6.

[90] *Crude [Oil] by Rail Accounts for More Than Half of East Coast Refinery Supply in February*, ENERGY INFO. ADMIN., May 5, 2015, https://www.eia.gov/todayinenergy/detail.cfm?id=21092.

[91] *Rail Deliveries of U.S. Oil Continue to Increase in 2014*, ENERGY INFO. ADMIN., Aug. 28, 2014, https://www.eia.gov/todayinenergy/detail.cfm?id=17751.

[92] Christopher J. Barr, *Unfinished Business: FERC's Evolving Standard for Capacity Rights on Oil Pipelines*, 32 ENERGY L.J. 563, 569 (2011).

[93] *Id.* at 569–70.

[94] Jeffrey O. Moreno & Thompson Hine, *The Hidden Issues in Crude-By-Rail Contracting*, OIL & GAS MONITOR, July 2, 2014, http://www.oilgasmonitor.com/hidden-issues-crude-rail-contracting/.

III. State Regulation of Energy Markets

As introduced in Section I, the federal government generally controls the transportation of energy products interstate—including the price that pipelines and electric transmission line operators may charge for use of transportation infrastructure. It also retains control over wholesale sales of electricity and natural gas, although it has chosen to largely deregulate this sector. However, states also play an important role in regulating energy markets. States regulate the generation of electricity for the purposes of selling that electricity to retail customers, and they regulate oil and natural gas production. States also have regulatory control over access and price for intrastate electric transmission lines as well as for intrastate oil and natural gas pipelines, and they regulate retail sales of electricity and natural gas and the distribution of electricity and natural gas from utilities to retail customers. This section focuses on state regulation of the intrastate transmission, distribution, and retail sale of electricity as a case study in state energy regulation, only briefly touching upon oil and natural gas.

A. State Regulation of Electricity Generation for Retail Sale

Although the federal government regulates the wholesale sales of energy from electricity generators to other utilities, states play a primary role in determining whether a power plant may be built at all—and thus whether a power plant may come online in order to produce wholesale or retail electricity, or both. Most U.S. states still treat the business of building and operating generating power plants as a business that is intertwined with two other activities—transmitting and distributing electricity—and they treat this business as a "natural monopoly." Thus, whenever an entity proposes to build a power plant within a state, the state must approve the construction of this plant by issuing a certificate of need or certificate of convenience and necessity to the plant. In some highly-regulated states, the state will not approve the proposed plant unless the developer can show that the majority of electricity produced by the plant is committed to be sold to entities that will sell electricity retail to in-state customers—and thus that the plant is "needed" to serve in-state customers. Thus, a company that proposes to build a power plant that might primarily produce wholesale electricity—the sales of which would be regulated by the federal government—can be blocked by a state agency, which retains "gatekeeping" authority over the market for constructing power plants. If a power plant would not serve in-state customers or provide wholesale electricity that would primarily be sold to in-state customers, or if the state believes that the plant will be too expensive, the state may deny the certificate of need.[95] This is true even for nuclear power plants, for which the *federal*, rather than the state, government must approve plant siting, design, and construction. States still may block nuclear power plants by denying them a certificate of need on the grounds that they are too expensive, and that the retail rates charged for sales from the plant to retail customers will be too high.[96]

Some states have taken a different path. They have determined that viewed independently, the business of building and operating a power plant is a competitive enterprise and need not be a natural monopoly. In these states that have "restructured" or at least partially restructured their electricity sectors, a developer of a proposed power plant need not first obtain a certificate establishing that the sale of power from the plant will be in the public interest or is "needed." Rather, the developer need only obtain a siting certificate from the state (in states that have centralized the power plant siting process) or a license to operate. In these restructured states, retail consumers may choose which generator they buy electricity from.

B. State Regulation of Intrastate Transmission Lines

Once a developer has built a power plant, the developer must transmit electricity from the site of generation to population centers. Chapter 1 discusses the fact that even for interstate electric transmission lines, it is primarily state law rather than federal law that governs whether and where a transmission line is built. But once built, the level of regulation of the use and charges associated with that line differ depending on whether the line is interstate or intrastate. Most transmission lines used for electricity transmission are interstate lines regulated by FERC (with the exception of siting).

Even many lines that are located wholly intrastate are considered to be "interstate" lines for legal purposes. For example, in the 1972 case *Federal Power Commission v. Florida Power & Light*,[97] Florida Power & Light (FPL) operated a network of power plants and transmission lines connecting the plants, and all of these lines were in the state of Florida.[98] However, FPL had transmission lines that were connected to another Florida utility, and this Florida utility had transmission lines that crossed the Florida-Georgia border. This utility sometimes sent electricity to Georgia and received electricity from across the border.[99] Therefore, because FPL's lines were

---

[95] *See infra* note 120 and accompanying text.

[96] *But see* Entergy Nuclear Vermont Yankee v. Shumlin, 733 F.3d 393 (2d. Cir. 2013) (in a market within an ISO, suggesting that a state will have a more difficult time justifying the denial of nuclear power plants).

[97] 404 U.S. 453 (1972).

[98] *Id.* at 456.

[99] *Id.* at 456–57.

connected to this utility, it was possible that electricity sent through FPL's wires to the utility would commingle with electricity from Georgia. The U.S. Supreme Court found that FPL's electricity was transmitted in interstate commerce and therefore was governed by what was then the Federal Power Commission (now FERC).[100]

Despite the relatively broad test for federal jurisdiction over transmission lines, some transmission lines remain wholly intrastate and thus are fully regulated by states. In fact, in Hawaii and Alaska, all transmission lines are regulated by these states because the transmission lines are isolated from other, out-of-state lines.[101] The same is generally true for Texas. Although a limited number of transmission lines cross the Texas border, the state's transmission grid is largely isolated, its electricity sales and transmission do not count as "interstate sales," and these sales and transmission are thus governed by the state through an organization called the Electric Reliability Council of Texas.[102]

Some states treat transmission lines as classic natural monopolies, in which one utility generates electricity, uses its own transmission lines to move electricity to population centers, and then distributes electricity to customers. These states have not forced intrastate transmission lines to be open access. Other states, however, have required the utilities that own and operate intrastate transmission lines to make these lines open access and to wheel electricity for other utilities. States that have fully restructured have required that utilities that used to be vertically integrated—meaning they were in the business of generating, transmitting, and distributing electricity—to functionally separate their generation business from transmission and distribution. In these states (with Texas being the primary example), there are now companies that are wholly in the business of transmitting and distributing electricity and are called transmission and distribution utilities, or TDUs.

Even with the functional separation of owning and operating wires from building and constructing power plants, transmission and distribution utilities are not fully competitive. This is because their wires are "bottlenecks." It would be inefficient to have thousands of competing companies building overlapping wires, so the state must limit and approve the number of lines built and operated. Therefore, even in states like Texas TDUs are still regulated entities. They must obtain certificates of need before building wires, and the state regulates the rates that they may charge of generators that use their

---

[100] *Id.* at 468–69.

[101] *Cf.* New York v. Fed. Energy Regulatory Comm'n, 535 U.S. 1, 7 (2002) (noting that the electric grids in these states are isolated).

[102] *See* Jared M. Fleisher, *ERCOT's Jurisdictional Status: A Legal History and Contemporary Appraisal*, 3 TEX. J. OIL GAS & ENERGY L. 4 (2008).

lines.[103] The state also requires TDUs to offer transmission lines on an open access basis.[104]

C. State Regulation of the Distribution of Electricity to Retail Customers

As introduced above, in states that still fully regulate the electricity sector, vertically integrated utilities are true natural monopolies that generate, transmit, and distribute electricity to retail customers and are in the business of measuring how much electricity customers use and billing customers for this use. In these states, customers do not typically get to choose the type of power plant that generates the electricity that flows to their home, business, or industrial facility. Instead, one utility has the exclusive authority to provide electricity to customers within a particular service territory, and customers within this service territory may only obtain electricity from this utility.[105] However, some vertically integrated utilities offer programs like "green buying" programs in which customers can choose to obtain electricity from renewable energy providers that sell to the utility.[106]

Some states have partially restructured this system, requiring vertically integrated utilities to divest their generation resources or at least functionally separate generation from transmission and distribution.[107] This causes generation to be a competitive industry. In some of these partially restructured states, retail customers still may not choose the generator from which they purchase electricity, but the utility that serves these retail customers often purchases power from various competitive generators rather than generating all of its own electricity.[108] In other partially restructured states, customers may choose the generator from which they receive electricity, but they are still required to use the vertically integrated utility as the entity that delivers the electricity to the home, business,

---

[103] *Transmission and Distribution Rates for Investor Owned Utilities*, PUB. UTIL. COMM'N OF TEX., https://www.puc.texas.gov/industry/electric/rates/TDR.aspx.

[104] *Id.*

[105] MUCHOW & MOGEL, *supra* note 28, at 52–23 to 52–25; REGULATORY ASSISTANCE PROJECT, ELECTRICITY REGULATION IN THE U.S.: A GUIDE, 25–26 (2011).

[106] RYAN WISER ET AL., LAWRENCE BERKELEY NAT'L LAB., CUSTOMER CHOICE AND GREEN POWER MARKETING: A CRITICAL REVIEW AND ANALYSIS OF EXPERIENCE TO DATE 3–4 (2000).

[107] *See supra* note 9.

[108] *See, e.g.*, ENERGY INFO. ADMIN., THE CHANGING STRUCTURE OF THE ELECTRIC POWER INDUSTRY 2000: AN UPDATE 86–87 (2000), http://webapp1.dlib.indiana.edu/virtual_disk_library/index.cgi/4265704/FID1578/pdf/electric/056200.pdf (describing Massachusetts, in which customers continue to receive distribution services from what was previously their vertically-integrated utility but may choose their supplier and may choose to receive either one electricity bill from their utility or bills that show the cost of electricity generation and the cost of distribution).

or industrial facility, meters the electricity, and bills the retail customer.

In fully restructured states, even the business of metering the electricity that is distributed to retail customers and billing those customers is a competitive enterprise that is no longer part of a vertically integrated activity. For example, in Texas, companies called retail electric providers (REPs) go out and find electricity and then approach retail customers to offer these customers a particular electricity package.[109] These REPs compete on a variety of levels. They offer retail customers different types of electricity—such as "green" packages, or packages that provide the cheapest possible electricity. They also offer billing on different cycles and fixed versus variable electricity rates.[110] Although REPs are competitive, they still must register and be licensed by the state of Texas before operating[111]—largely for purposes of consumer protection.

D. Oil and Natural Gas: State Regulation of Intrastate Pipelines, Distribution Lines, and Retail Sales

Just as states regulate intrastate transmission and distribution of electricity and retail electricity sales, states also regulate the construction and siting of intrastate oil and natural gas pipelines and distribution lines as well as retail sales of natural gas from "local distribution companies" to homes and businesses.[112] Many states require natural gas and oil pipelines and distribution line companies to obtain a certificate of public convenience and necessity or certificate of need from the state public service commission, corporation commission, or similar agency prior to building these lines.[113] Many also require siting certificates[114] and grant eminent domain authority to these pipeline companies.[115] For more discussion of intrastate pipeline siting and eminent domain authority, see Chapter 1. Further, many states require companies that sell natural gas retail to customers to charge a regulated rate—a rate set by a

---

[109] Certification and Licensing, *REP–Retail Electric Providers Certification and Licensing*, PUB. UTIL. COMM'N OF TEX., https://www.puc.texas.gov/industry/electric/business/rep/rep.aspx.

[110] *Plan Results*, POWERTOCHOOSE.ORG, http://www.powertochoose.org/en-us/Plan/Results (entering "75201"—a Dallas zip code, and then "view results" brings up more than 300 options for retail customers).

[111] *See supra* note 109.

[112] U.S. GOV'T ACCOUNTABILITY OFF., PIPELINE PERMITTING: INTERSTATE AND INTRASTATE NATURAL GAS PERMITTING PROCESSES INCLUDE MULTIPLE STEPS, AND TIME FRAMES VARY 22, http://www.gao.gov/assets/660/652225.pdf.

[113] *See, e.g.*, FLA. STAT. ANN. § 403.9422 (West 2016); Alexandra B. Klass, *Future-proofing Energy Transport Law*, 94 WASH. U. L. REV. ___ (forthcoming 2017), *available at* http://papers.ssrn.com/sol3/papers.cfm?abstract_id=2748905.

[114] U.S. GOV'T ACCOUNTABILITY OFF., *supra* note 112, at 23–24.

[115] Klass, *supra* note 113, at 8.

state agency that is designed to give the company the opportunity to recover the cost of building and operating the infrastructure designed to deliver the natural gas and to obtain a rate of return from investors.[116] Other states, however, have required natural gas companies to "unbundle" (separate) the business of transporting and distributing gas from the business of purchasing natural gas, thus allowing customers to choose their natural gas supplier.[117] In these states, the rates that companies charge for use of intrastate natural gas pipelines remains regulated, but the price of retail natural gas is unregulated.[118]

E. Conflicts Between State and Federal Jurisdiction

Because both the federal government and states have important authority over certain aspects of energy markets, and retail and wholesale markets are inherently intertwined, conflicts between these two levels of government often arise. For example, say that a large utility owns a power plant. The utility produces electricity to sell to other utilities wholesale (under a FERC-approved market rate) and also sells some of its electricity retail (directly to customers, under a PUC-approved rate). Further, the utility purchases wholesale electricity from other utilities and sells this electricity retail. Say that a state PUC believes that the utility is paying too much for the wholesale electricity that it purchases and then sells to retail customers, passing on the costs to retail customers through the retail rate approved by the PUC. When the PUC approves the retail rate, may it "pick apart" the wholesale rate and decide that the wholesale rate was improperly calculated by FERC, and thus that the utility should only recover a portion of that rate from retail customers? The short answer is "no." Just as the filed rate doctrine provides that a rate "filed" with and approved by FERC may only be challenged through FERC—not the courts—it also provides that states may not interfere with these FERC-approved wholesale rates, even if they are market rates. In other words, "[t]he filed rate doctrine requires 'that interstate power rates filed with FERC or fixed by FERC must be given binding effect by state utility commissions determining intrastate rates.'"[119] However, states have other ways of modifying utilities' plans to purchase wholesale power. When the utility goes to the PUC requesting approval to purchase

---

[116] U.S. ENERGY INFO. ADMIN., DISTRIBUTION OF NATURAL GAS: THE FINAL STEP IN THE TRANSMISSION PROCESS 3 (2008), https://www.eia.gov/pub/oil_gas/natural_gas/feature_articles/2008/ldc2008/ldc2008.pdf.

[117] *See Natural Gas Residential Choice Programs*, U.S. ENERGY INFO. ADMIN., May 17, 2010, http://www.eia.gov/oil_gas/natural_gas/restructure/restructure.html.

[118] *See, e.g.*, *Retail Unbundling—Colorado*, U.S. ENERGY INFO. ADMIN., http://www.eia.gov/oil_gas/natural_gas/restructure/state/co.html.

[119] Entergy Louisiana, Inc. v. Louisiana Pub. Serv. Comm'n, 539 U.S. 39, 47 (2003) (quoting Nantahala Power & Light Co. v. Thornburg, 476 U.S. 953, 962 (1986)).

wholesale power to sell to its retail customers, the PUC could simply deny the proposed wholesale purchase altogether. Furthermore, when utilities propose to build power plants that will provide both wholesale and retail power, state PUCs can deny the certificate of need for the plant.[120]

Another conflict that has arisen between state and federal jurisdiction involves demand response resources discussed in more detail in Chapter 8. Demand response is the practice of reducing electricity use during periods of peak electricity demand; instead of having to draw from expensive peak generation sources, the entity scheduling electricity flow can instead rely on this energy "non-use" to avoid the situation of more electricity being demanded than is available for power plants. FERC issued an order providing that electricity customers—including retail customers like home owners and business owners—could bid their electricity non-use (demand response) into wholesale markets and receive compensation comparable to the compensation that generators received in those markets. Retail customers' ability to bid their energy non-use into wholesale markets was facilitated by companies called "aggregators," which bundle together various businesses' or home owners' promises to curtail electricity use during periods of peak demand and bid this valuable, bundled product into wholesale markets. Generators challenged this FERC order, arguing that FERC lacked jurisdiction over retail demand response bid into wholesale markets. In *FERC v. Electric Power Supply Association*, the U.S. Supreme Court determined that FERC does have jurisdiction over these demand response resources because "[t]he Federal Power Act . . . authorizes the Federal Energy Regulatory Commission . . . to regulate 'the sale of electric energy at wholesale in interstate commerce,' including both wholesale electricity rates and any rule or practice 'affecting' such rates."[121] The Court concluded that retail demand response resources that are bid into wholesale markets directly affect wholesale rates, and thus are subject to FERC jurisdiction, because FERC's rules for bidding these resources into wholesale markets allow operators of wholesale markets to "accept such bids if and only if they bring down the wholesale rate by displacing higher-priced generation."[122] Nevertheless, both FERC and the Court also held that states could put limits on the ability of retail customers to participate

---

[120] *See, e.g.*, Tampa Elec. Co. v. Garcia, 767 So.2d 428, 435–36 (Fla. 2000) (finding that only plants that will commit the majority of their electricity to in-state customers may obtain a certificate of need).

[121] 136 S. Ct. 760, 766 (2016) (quoting 16 U.S.C. §§ 824(b), 824e(a)).

[122] *Id.* at 774.

in wholesale markets for demand response, and some states have done so.[123]

A final wholesale-retail dispute that raises state versus federal jurisdictional issues involves generation capacity. "Capacity" is a commitment by a generator that in the future, it will have a power plant available to supply electricity that will be necessary to meet increased demand. Entities that operate transmission lines, including RTOs and ISOs, must ensure that they will have adequate generation capacity to call on when electricity is being demanded; inadequate capacity causes an "undervoltage"—meaning not enough electricity is flowing through the grid—and can cause blackouts or brownouts. In *Hughes v. Talen Energy*,[124] the U.S. Supreme Court addressed a situation in which an RTO called PJM, which controls transmission lines that run from the Midwest through the Midatlantic and parts of the Northeast, runs a "capacity auction." As explained by the court:

> The PJM capacity auction functions as follows. PJM predicts electricity demand three years ahead of time, and assigns a share of that demand to each participating LSE [load-serving entity, which is a generator]. Owners of capacity to produce electricity in three years' time bid to sell that capacity to PJM at proposed rates. PJM accepts bids, beginning with the lowest proposed rate, until it has purchased enough capacity to satisfy projected demand. No matter what rate they listed in their original bids, all accepted capacity sellers receive the highest accepted rate, which is called the "clearing price." LSEs then must purchase from PJM, at the clearing price, enough electricity to satisfy their PJM-assigned share of overall projected demand. The capacity auction serves to identify need for new generation: A high clearing price in the capacity auction encourages new generators to enter the market, increasing supply and thereby lowering the clearing price in same-day and next-day auctions three years' hence; a low clearing price discourages new entry and encourages retirement of existing high-cost generators.[125]

Within this market, all generators whose bids are accepted receive a price equal to the last price offered by the last generator necessary to satisfy the quota. (For example, say generator A bids 1 MW of capacity for $1/MW, generator B bids 2 MW of capacity for $2/MW, and generator C bids 3 MW of electricity for $3/MW. PJM

---

[123] *Id.* at 779–80.
[124] 136 S. Ct. 1288 (2016).
[125] *Id.* at 1293.

needs 4 MW of capacity, and therefore needs all of generator A's and B's capacity and 1 MW from generator C to satisfy its quota. Generators A, B, and C would all receive $3 for the MW they provide.) To avoid gaming of the market, PJM—as required by FERC rules—sets a minimum price that all participants in the auction have to initially offer—even those generators that are regulated by states.[126] (Recall that states have the power to deny the construction of new plants, including plants that will primarily generate wholesale power, if the states think that those plants are not needed. States also may prevent utilities from recovering certain capital, operational, or electricity purchase costs through retail rates if the states believe that those rates are unjust or unreasonable—provided that those state decisions do not "pick apart" or interfere with FERC-approved wholesale rates.)

States within PJM, such as Maryland and New Jersey, were concerned that PJM markets were not adequately incentivizing the construction of new generation capacity; they were concerned that there would be inadequate amounts of electricity generated for retail customers who needed the electricity.[127] Thus, states like Maryland and New Jersey began offering incentive payments through fixed contracts to certain generators that promised to build new capacity.[128] (The case only addresses Maryland's program but mentions New Jersey's similar program.) These generators were guaranteed the contract price rather than the price they would receive through the PJM capacity auction.[129] Competing power plants believed that these state incentive payments were affecting the price at which generators bid their capacity into wholesale markets, and were thus interfering with the wholesale market prices.[130] In 2016, the U.S. Supreme Court found that the Federal Power Act preempted Maryland's payment scheme because it was indeed interfering with wholesale rates—a practice prohibited under the filed rate doctrine.[131]

This Chapter has shown that energy markets vary dramatically depending on the physical stage of energy involved. (Electric transmission markets are very different from oil and gas production and electricity generation markets, for example.) They also vary from the federal to the state level, although states in many cases have followed FERC's lead and have restructured certain markets.

---

[126] *Id.* at 1294.
[127] *Id.*
[128] *Id.* at 1294–95, n.4.
[129] *Id.* at 1295.
[130] *Id.* at 1296.
[131] *Id.* at 1298–99.

Chapter 7

ENERGY IMPORTS AND EXPORTS

I. Introduction

Just as Chapter 6 showed that the federal government regulates many aspects of domestic energy markets, the federal government also tightly controls certain exports of energy products from the United States. This is in part because policymakers tend to treat energy differently than they treat other products. Many policymakers, while supporting domestic production of goods generally, tend to accept the fact that U.S. citizens and businesses commonly rely on imports for many of the goods that we buy in this country—say, bananas, or consumer appliances. But these same policymakers also believe that it is important that we produce our *energy* products domestically, to the greatest extent possible. Those who subscribe to this policy typically cite to the concept of energy security—the idea that we should avoid purchasing too many fuels from foreign countries, particularly from nations that do not share our democratic values or that have other domestic disputes with us. Thus, pointing to energy security concerns, the United States has historically limited exports of domestically produced fuels.

Relatively complex regulation of energy exports is also a product of history. In 1973, members of the Organization of Arab Petroleum Exporting Countries (OPEC) agreed to stop exporting oil to the United States and several other countries as a result of conflicts with Israel, leading to U.S. fuel shortages and long lines at gas pumps.[1] This spurred the United States to focus more on incentivizing domestic energy production and the use of domestic fuels, and it left a lasting mark on U.S. energy policy.

Despite the focus on energy security and concerns about overreliance on fuel imports, fuels have long been a global product, just like many products that are imported or exported by countries based on global supply and demand forces and price differentials. And the United States has long been a major importer of fuels. However, the amount of fuels exported from the United States has recently increased due to the boom in "slick water" fracturing that occurred throughout the United States starting in the mid-to late-2000s. This boom created a glut of oil, leading many oil producers to send their product to refineries, from which the oil was then exported. The oil glut also incentivized oil producers to lobby for

---

[1] Luis E. Cuervo, *OPEC from Myth to Reality*, 30 HOUS. J. INT'L L. 433, 594, n.790 (2008).

changes in U.S. export policy, which previously prohibited most exports of crude oil from the United States and only allowed exports of refined oil. Oil producers got their way in 2016, when Congress lifted the ban.[2] Exports of natural gas still require federal approval, but these exports, too, have increased. Although U.S. exports of fuels have grown in recent years, fuel companies in the United States continue to import large quantities of oil and gas for reasons of convenience (some users of oil and gas are located near convenient import points) and global market forces. This Chapter briefly discusses the history of fuel imports in the United States and then focuses on the regulation of oil exports and natural gas import and export. This Chapter does not address coal imports or exports because these transactions are not regulated, aside from the regulation of the siting and construction of coal export terminals, which we discuss in Chapter 1.

II. Energy Import and Export Trends

Although the United States has abundant coal, natural gas, and oil reserves, fuel purchasers have long imported fuel into the United States. As introduced above, this is due to global prices as well as supply and demand. Additionally, for certain periods of time it appeared that we were running out of easily-accessible reserves of oil and gas. The U.S. government tried to incentivize oil and gas production companies to go out and drill for more difficult-to-access unconventional reserves,[3] but these efforts were not always entirely successful.

With respect to natural gas, fuel purchasers in the United States imported relatively steady but low amounts of natural gas in the 1970s, with a temporary uptick in 1979. In 1975 we imported approximately 953 trillion cubic feet of natural gas, then a higher amount (1.3 million cubic feet) in 1979, and then a lower amount again in 1980 (approximately 985 trillion cubic feet).[4] After 1986, imports then steadily rose, peaking at approximately 4.6 trillion cubic feet of natural gas imported in 2005.[5] Then, due largely to the slick water fracturing boom that gave gas producers access to natural

---

[2] Consolidated Appropriations Act, 2016, H.R. 2029.

[3] For example, when the Federal Energy Regulatory Commission still regulated the price of natural gas that was sold "at the wellhead" (just as it came out of the well) to a buyer that eventually sent the gas interstate, Congress through the Natural Gas Policy Act of 1978 allowed some prices to rise above the regulated rate. This included natural gas extracted from more difficult-to-access unconventional formations, such as "deep gas"—gas from deposits trapped deep underground. *See* Paul W. MacAvoy et al., *Is Competitive Entry Free? Bypass and Partial Deregulation in Natural Gas Markets*, 6 YALE J. ON REG. 209, 218–19 (1989).

[4] *Natural Gas: Natural Gas Imports*, U.S. ENERGY INFO. ADMIN., https://www.eia.gov/dnav/ng/hist/n9100us2a.htm.

[5] *Id.*

gas trapped tightly within shale and tight sandstone rock formations underground, imports have generally steadily declined since 2005, dropping to 2.7 trillion cubic feet in 2015.[6] During this period, U.S. fuel importers purchased less gas (which is primarily imported from Canada), and U.S. fuel producers exported far more gas to Mexico. For example, exports to Mexico increased by 10 percent between 2013 and 2014.[7] These types of trends caused U.S. net imports of natural gas to steadily decline since 2007.[8]

The history of imports of crude oil and petroleum products to the United States follows a trend that is similar to natural gas. Between 1980 and 1985 oil imports remained somewhat flat and steady (although experiencing rises and dips annually), with imports ranging from 6.909 million barrels of oil per day imported in 1980 to 5.067 million barrels of oil per day imported in 1985.[9] Oil imports then generally rose steadily, peaking at 13.714 million barrels of crude oil per day imported in 2005.[10] Then, again due largely to the slick water fracturing boom, oil imports began to decline relatively steadily, with imports totaling approximately 9.401 million barrels in 2015.[11]

The recent boom in domestic oil and gas production, which has caused net imports of these products to drop, has driven U.S. producers to seek more export opportunities. However, producers that wish to export certain types of oil and gas historically faced regulatory hurdles, as discussed in the following part.

III. U.S. Regulation of Energy Imports and Exports

U.S. regulation of oil and gas imports and exports depends on the type of fuel being exported. For example, U.S. producers have long been allowed to import any type of oil and gas and to export *refined* oil, but crude oil exports from the United States were generally banned, with exceptions. The Bureau of Industry and Security (BIS) within the Department of Commerce had to approve

---

[6] *Id.*

[7] *Natural Gas: U.S. Natural Gas Imports and Exports 2014*, U.S. ENERGY INFO. ADMIN., https://www.eia.gov/naturalgas/importsexports/annual/.

[8] *Id.*

[9] *Petroleum & Other Liquids: U.S. Imports of Crude Oil and Petroleum Products*, U.S. ENERGY INFO. ADMIN., http://www.eia.gov/dnav/pet/hist/LeafHandler.ashx?n=PET&s=MTTIMUS2&f=A

[10] *Id.*

[11] *Id.*

oil exports.[12] As introduced above, in 2016 Congress lifted the export ban,[13] thus ending the need for BIS approval.

In the case of natural gas, most exports—with the exception of exports of natural gas via pipeline to Mexico—take the form of liquefied natural gas (LNG) exports. Liquefied natural gas is gas that has been cooled to a very low temperature. This liquefaction allows for more efficient transport of natural gas on various types of vessels overseas. After the LNG is transported to a foreign port, it is regasified and transported by pipeline for use in another country. All natural gas exports, whether they are exports of the actual gas or LNG, require approval by the U.S. Department of Energy (DOE)[14] while the terminals used to import or export the gas require approval by the Federal Energy Regulatory Commission (FERC).[15]

A. Crude Oil Exports

From 1975 through 2015, oil imports were largely unrestricted (aside from occasional foreign policy changes that limited U.S. access to oil), but it was very difficult for U.S. oil producers to export crude (unrefined) oil. This was due to the Energy Policy Conservation Act of 1975 (EPCA), which required the President to exercise authority provided within the Act to "promulgate a rule prohibiting the export of crude oil and natural gas produced in the United States," with certain exemptions permitted.[16] The resulting rules banned most crude oil exports absent a BIS finding that the exports were in the "national interest."[17] Exceptions to this requirement for an individualized determination of "national interest" included exports that were permitted if the BIS issued a license for the particular export request, including, inter alia, exports to Canada; exports from certain parts of Alaska; and exports of certain types of oil from California.[18] Certain other crude oil exports did not even require a license. These included, among others, exports of small amounts of oil for sampling or testing, exports of oil that flowed through the

---

[12] *See* PHILLIP BROWN ET AL., CONG. RESEARCH SERV., U.S. CRUDE OIL EXPORT POLICY: BACKGROUND AND CONSIDERATIONS 6–7 (Mar. 26, 2014) (describing the source of BIS authority over crude oil exports).

[13] Consolidated Appropriations Act, 2016, H.R. 2029.

[14] 15 U.S.C. § 717b (West 2016) (requiring approval of gas imports and exports by what was previously the Federal Power Commission); 42 U.S.C. § 7151 (West 2016, originally enacted in 1977) (DOE Organization Act, transferring certain Federal Power Commission authority to the DOE).

[15] 15 U.S.C. § 717b(a) (West 2016).

[16] Pub. L. no. 94–163 (1975).

[17] 42 U.S.C. § 6212(b)(1) (West 2016). For a history of the ban and resulting regulations, *see* Sam Andre, Note, *Striking Before the Well Goes Dry: Exploring If and How the United States Ban on Crude Oil Exports Should be Lifted to Exploit the American Oil Boom*, 100 MINN. L. REV. 763, 763–64, 766–68 (2015).

[18] 15 C.F.R. § 754.2(b) (West 2016).

Trans-Alaska pipeline, and exports of oil from the United States' strategic petroleum reserve.[19] For any other type of crude oil export, an oil producer had to apply to the BIS for an individual determination that export was in the "national interest" and consistent with the EPCA.[20] The national interest determination required the BIS to consider whether the exports would result in imports equal to the same or larger amount of oil being exported; whether the export contracts could be terminated in the event of U.S. oil supply emergencies; and whether the applicant could show "compelling economic or technological reasons" beyond the applicant's control that prevented U.S. marketing of the oil.[21] Alternatively, producers had to show that the oil was not in fact "crude."[22]

Before the United States lifted the export ban, there was some hope that the BIS would start to loosen its definition of crude oil. Typically, for oil to be considered "not crude," and thus not to be subjected to the case-by-case "national interest" determination required for export (if the oil did not fall within an exemption), it had to be distilled at a refinery. Specifically, it had to be heated up and sent through a distillation tower, where the oil experienced different temperatures at different points throughout the tower, was condensed into different products at these points, and flowed out of the tower to different collection tanks and pipes. In 2014, two Texas oil production companies applied to the BIS for a determination that "condensate"—very light oil produced from their wells—was not crude because it had been treated in a "stabilization unit," which is different from a distillation tower but still treats the condensate in basic ways. In this type of unit, the condensate is heated up, and the lighter, less stable components of the condensate like methane, ethane, propane, and butane vaporize and are separated off, leaving behind the heavier liquids.[23] The BIS issued confidential letter rulings concluding that the condensate treated in the stabilization unit was in fact not crude oil and could be exported without BIS approval.[24] But the BIS made clear that these letters were not indicative of a broader BIS policy shift to narrow its definition of crude oil and require fewer BIS approvals of crude export; rather,

---

[19] BROWN ET AL., *supra* note 12, at 5.

[20] 15 C.F.R. § 754.2(b) (West 2016).

[21] 15 C.F.R. § 754.2(b)(2)(i)(C) (West 2016).

[22] For a longer, similar description of the previous export ban and exceptions to it see JOEL B. EISEN ET AL., ENERGY, ECONOMICS, AND THE ENVIRONMENT: CASES AND MATERIALS ch. 14 (4th ed. 2015).

[23] *Condensate Stabilizer Unit*, EXTERRAN, http://www.exterran.com/Content/Docs/Products/Condensate%20Stabilizer-English-A4.pdf.

[24] *Oil Exports*, TRANSP. & LOGISTICS INT'L MAGAZINE, http://www.tlimagazine.com/sections/columns/2137-oil-exports.

any producers not falling under an exemption would continue to have to obtain case-by-case "national interest" approvals.[25]

With little hope that the BIS would allow more exports, U.S. oil producers scrambled to find another way to ease the process of exporting crude oil from the United States. This was particularly important because much of the oil produced from the slick water fracturing glut could be more easily treated in European refineries.[26] Thus, oil producers lobbied Congress for a change, and Congress ultimately ended the export ban in the 2016 budget,[27] providing that "no official of the Federal Government shall impose or enforce any restriction on the export of crude oil,"[28] with certain limited exceptions. This involved a policy compromise in which Congress also extended tax credits for U.S. renewable energy.[29]

B. Liquefied Natural Gas Exports

Congress did not lift the export restrictions on natural gas when it ended the crude oil export ban. Further, unlike with oil, natural gas imports *and* exports require regulatory approval. Thus, any natural gas producer, company, or marketer wishing to export natural gas from the United States or import gas—typically in liquefied form, unless the natural gas is sent by pipeline to a nearby country like Mexico (which is an increasingly common practice[30])—must obtain approval from the DOE.[31] Many natural gas producers are seeking to export gas because of price differentials in global markets. Although these differentials have declined in recent years, natural gas is still more expensive in parts of Asia and other global regions than it is in the United States,[32] thus creating potential profits for U.S. producers who export gas. Because requests for

---

[25] *Id.*; *see also* EISEN ET AL., *supra* note 22, ch. 14, for a similar description of the BIS letters.

[26] EISEN ET AL., *supra* note 22, ch. 14; Steven Mufson, *Did the Obama Administration Just Lift the Ban on U.S. Crude Oil Exports?*, WALL ST. J., June 25, 2014.

[27] Consolidated Appropriations Act, 2016, H.R. 2029.

[28] *Id.* at Div. O, Title I.

[29] *Id.* at Div. P, Title III.

[30] Christine Buurma, *Natural Gas: Mexican Pipeline to Boost U.S. Gas Exports 20 Percent*, 88 DAILY ENVIRONMENT REPORT A–18, May 6, 2016.

[31] 15 U.S.C. § 717b (West 2016).

[32] *See* U.S. ENERGY INFO. ADMIN., WORLD LNG ESTIMATED FEBRUARY 2016 LANDED PRICES, http://www.ferc.gov/market-oversight/mkt-gas/overview/ngas-ovr-lng-wld-pr-est.pdf (showing prices ranging from $1.37 in the United States to $4.57 in South America); MICHAEL RATNER ET AL., CONG. RESEARCH SERV., U.S. NATURAL GAS EXPORTS: NEW OPPORTUNITIES, UNCERTAIN OUTCOMES 3 (Jan. 28, 2015), https://www.fas.org/sgp/crs/misc/R42074.pdf (noting that "[l]ow U.S. natural gas prices (especially in the Gulf of Mexico) relative to other international markets have spurred interest in exporting U.S. natural gas").

natural gas export are now more common than requests for import,[33] (and because the procedures required for obtaining approval of natural gas import and export are nearly identical), this section will focus on export requirements.

Similar to oil (prior to Congress's lifting the crude oil export ban), the type of natural gas export approval required from the DOE varies. Due to a Congressional amendment to the natural gas export limitations in 1992,[34] if the proposed export is to a country with which the United States has a free trade agreement (FTA) specific to natural gas, in which the country agrees to "national treatment" of natural gas, the export is automatically deemed to be in the public interest and may immediately occur.[35] "National treatment" of natural gas means that the countries with which we have an FTA "may not discriminate between 'like products' that are imported and those domestically produced."[36] In other words, these countries may not favor domestic natural gas over imported natural gas. The United States has these types of natural gas FTAs with approximately eighteen countries.[37]

For exports to non-FTA countries, the DOE must find that the proposed export of natural gas is consistent with the public interest.[38] Regardless of whether the exporter is proposing to send gas to an FTA or non-FTA country, an exporter proposing to export gas over the long-term (as opposed to one isolated export transaction) must

---

[33] The shift from requests for LNG import to LNG export has also been accompanied by requests to construct LNG export terminals. *See* Alexandra B. Klass, *Future-proofing Energy Transport Law*, 94 WASH. U. L. REV. (forthcoming 2017), *available at* http://papers.ssrn.com/sol3/papers.cfm?abstract_id=2748905 (noting the shift from building natural gas import terminals to export terminals "in a span of less than ten years"); RATNER ET AL., *supra* note 32, at 2 ("The United States currently has LNG import capacity of almost 14 billion cubic feet per day (bcfd) or over 5 trillion cubic feet (Tcf) per year. However, higher domestic production has made imports largely unnecessary, leaving existing import capacity mostly idle.").

[34] Energy Policy Act of 1992, Pub. L. no. 102–486, § 201 (amending section 3 of the Natural Gas Act to provide that "the exportation of natural gas to a nation with which there is in effect a free trade agreement requiring national treatment for trade in natural gas, shall be deemed to be consistent with the public interest, and applications for such importation or exportation shall be granted without modification or delay").

[35] Donald F. Santa, Jr. & Patricia J. Beneke, *Federal Natural Gas Policy and the Energy Policy Act of 1992*, 14 ENERGY L.J. 1, 21 (1993) (noting that "the approval process for the export of domestic gas to FTA countries . . . [was] purely automatic" following the Energy Policy Act of 1992).

[36] Adam Eldean, *Can The United States Control its Natural Gas?: International Trade Implications of Restrictions on Liquefied Natural Gas Exports*, 54 NAT. RESOURCES J. 439, 454 (2014).

[37] *How to Obtain Authorization to Import and/or Export Natural Gas and LNG*, ENERGY.GOV, OFFICE OF FOSSIL ENERGY, http://energy.gov/fe/services/natural-gas-regulation/how-obtain-authorization-import-andor-export-natural-gas-and-lng#LNG (listing FTA countries as of 2012).

[38] 15 U.S.C. § 717b (West 2016).

submit an application to the DOE that contains various data points, including, for example, the "exact legal name" of the company that proposes to export gas and the contact information for the individual from the company that will serve as the DOE contact person; the proposed start date of export and volume of gas to be exported; the company that will supply the natural gas proposed to be exported; the point at which the gas will be exported; the geographic market or markets to which the gas will be sent; the name of the company that will transport the gas overseas; and "the major provisions of the gas purchase or sales contract."[39]

One recent application to export relatively small amounts of LNG from Jacksonville, Florida to FTA and non-FTA countries provides a helpful practical example. The company, Eagle LNG Partners Jacksonville LLC (Eagle LNG), submitted an application to the DOE through its law firm proposing to export approximately 49.8 billion cubic feet of LNG annually "for a period of twenty (20) years, commencing on the earlier of the date of first export or five years from the date of the final order granting export authorization"[40] (with the date projected to be "the fourth quarter of 2018."[41] The proposed point of export would be "at a site on the St. Johns River" in Jacksonville, and the supplier of the natural gas to be exported would be a "local utility" that purchases natural gas from a variety of producers.[42] Eagle LNG would receive natural gas from the utility, liquefy and store it, and "periodically load LNG onto ocean-going LNG carrier vessels for export."[43] Eagle LNG would also load the LNG into "trucks, containers and marine vessels" to be transported domestically.[44] The physical LNG export facility would take up approximately 54 acres of a 194-acre lot, and the siting and construction of the facility would be subject to FERC approval[45] (a process that we discuss in Chapter 1). In its application, Eagle LNG requested authorization to export its LNG to any country with which the United States currently or will in the future have an FTA, and to any non-FTA country.[46] In its application, the company noted that it had not yet entered into specific contracts for the sale of its LNG, in part because it needed authorization from the DOE before finalizing

---

[39] *How to Obtain Authorization to Import and/or Export Natural Gas or LNG*, supra note 37.

[40] *Eagle LNG Partners Jacksonville LLC*, Docket No. 16-15-LNG Application for Long-Term Authorization to Export Liquefied Natural Gas to Both FTA and Non-FTA Countries, Jan. 27, 2016, http://energy.gov/sites/prod/files/2016/02/f29/16-15-LNG.pdf.

[41] *Id.* at 4.

[42] *Id.* at 1.

[43] *Id.*

[44] *Id.* at 2.

[45] *Id.* at 3.

[46] *Id.* at 4.

contracts, but that it anticipated entering into "LNG Sales and Purchase Agreements ('SPAs')."[47]

After receiving the application, the DOE goes through the complex process of determining whether the proposed export is in the public interest. To apply the public interest standard the DOE investigates the economic, international, and environmental impacts of the export and its impacts on the security of the domestic natural gas supply.[48] To assist in its individualized public interest determinations for proposed natural gas exports, the DOE had two studies prepared regarding the economic impacts of exports, one by the U.S. Energy Information Administration[49] and the other by NERA Economic Consulting.[50] Further, although FERC takes the lead on preparing the Environmental Impact Statement (EIS) under the National Environmental Policy Act when natural gas exports and an associated export terminal are proposed, the DOE serves as a "cooperating agency in preparing the EIS."[51] The DOE previously conditionally approved natural gas exports while awaiting the completion of FERC's EIS, but in August 2014 it announced that it would change its procedures, summarizing its new procedures as follows:

> The U.S. Department of Energy . . . will act on applications to export liquefied natural gas . . . from the lower-48 states to countries with which the United States does not have a free trade agreement requiring national treatment for natural gas only after completing the review required by the National Environmental Policy Act (NEPA), suspending its practice of issuing conditional decisions prior to final authorization decisions.[52]

---

[47] *Id.* at 5.

[48] DOE Order No. 1473, Order Extending Authorization to Export Liquefied Natural Gas from Alaska (1999), https://www.ferc.gov/industries/gas/indus-act/angtp/doe1473.pdf; 1984 Policy Guidelines, 49 Fed. Reg. 6684 (Feb. 22, 1984).

[49] U.S. ENERGY INFO. ADMIN., EFFECT OF INCREASED NATURAL GAS EXPORTS ON DOMESTIC ENERGY MARKETS (Jan. 2012), http://energy.gov/sites/prod/files/2013/04/f0/fe_eia_lng.pdf.

[50] NERA ECONOMIC CONSULTING, MACROECONOMIC IMPACTS OF LNG EXPORTS FROM THE UNITED STATES (Dec. 2012), http://energy.gov/sites/prod/files/2013/04/f0/nera_lng_report.pdf.

[51] EIS–0501: Golden Pass LNG Export Project; Texas and Louisiana, ENERGY.GOV, http://energy.gov/nepa/eis-0501-golden-pass-lng-export-project-texas-and-louisiana; *see also* DEP'T OF ENERGY, PROCEDURES FOR LIQUEFIED NATURAL GAS EXPORT DECISIONS: FINAL REVISED PROCEDURES 2 (Aug. 15, 2014), http://www.energy.gov/sites/prod/files/2014/08/f18/Fed.%20Reg.%20Notice%20Procedures.pdf ("Typically, the agency responsible for permitting the export facility serves as the lead agency in the NEPA review process and DOE serves as a cooperating agency within the meaning of the Council on Environmental Quality's (CEQ) regulations." CEQ is the agency that directs other federal agencies how to comply with NEPA.).

[52] DEP'T OF ENERGY, *supra* note 51, at 1.

Thus, export approvals can sometimes be substantially delayed while awaiting the completion of the EIS. When FERC, DOE, and other agencies complete the EIS, they have chosen to consider (broadly speaking) impacts beyond the export terminal itself even though they do not believe that they are required to do so. Specifically, the agencies have repeatedly asserted that considering how export might induce the drilling and fracturing of more gas wells and increase various pollutant and greenhouse gas emissions from those wells is beyond the required scope of their analysis.[53] Nevertheless, the DOE has had another federal agency and outside experts prepare a report on "life cycle" greenhouse gas emissions associated with gas exports.[54]

This analysis looked at greenhouse gas emissions under three different natural gas scenarios, including: (1) natural gas produced from drilling and fracturing in Pennsylvania's Marcellus Shale, "transported by pipeline to an LNG facility where it is compressed and loaded onto an LNG tanker," transported to an Asian or European port, re-gasified, and "transported to a natural gas plant," where the gas would be burned to produce electricity, and the electricity would be transported through electric transmission lines and distribution lines to consumers; (2) scenario 1, except with gas being produced in several U.S. regions, not just the Marcellus Shale, and drilled using "conventional" methods (not fracturing); and (3) natural gas being produced in and near Siberia using conventional methods and "transported by pipeline to a power plant in Europe or Asia."[55] In all three scenarios, the report assumed that seven percent of electricity transported to consumers was lost.[56] (Some electricity is lost in the form of heat when transported long distances through transmission lines.)

The report concluded that "[t]he liquefaction, ocean transport, and regasification of natural gas are energy intensive activities with significant GHG [greenhouse gas] emissions, accounting for 17.5

---

[53] 148 FERC ¶ 61,244, Dominion Cove Point LNG, LP, Order Granting Section 3 and Section 7 Authorizations 76 (Sept. 29, 2014), https://www.ferc.gov/Calendar Files/20140929192603-CP13-113-000.pdf (noting that the environmental assessment for the project "explains that a specific analysis of Marcellus shale upstream facilities is outside the scope of this analysis because the exact location, scale, and timing of future facilities are unknown"); 151 FERC ¶ 61,095, Dominion Cove Point LNG, LP, Order Denying Rehearing and Stay 23 (May 4, 2015), http://www.ferc.gov/CalendarFiles/20150504181112-CP13-113-001.pdf ("[The future development of upstream production is speculative and not reasonably foreseeable. DOE acknowledges that its life cycle analysis . . . report goes beyond NEPA requirements and states that DOE cannot meaningfully analyze specific upstream impacts.").

[54] TIMOTHY J. SKONE, ET AL., U.S. DEP'T OF ENERGY, NAT'L ENERGY TECH. LAB., LIFE CYCLE GREENHOUSE GAS PERSPECTIVE ON EXPORTING LIQUEFIED NATURAL GAS FROM THE UNITED STATES (2014).

[55] Id. at 2–4.

[56] Id. at 4.

percent of the cradle-to-grave emissions" in certain production and export scenarios.[57] Emissions from "natural gas extraction, processing, and transport activities in the exporting country, in turn" contribute to "16.0 percent of the cradle-to-grave emissions."[58] So natural gas extracted from and burned in power plants within the United States has lower life-cycle emissions than natural gas produced in other countries and burned in other countries, as would be expected. However, natural gas produced in different countries has different methane emissions, which is important to note. Methane is a more potent greenhouse gas than carbon dioxide and remains in the atmosphere for a shorter time than CO_2.[59] According to the study commissioned by the DOE, methane emissions account for 13.8 percent of total lifecycle greenhouse gas emissions for gas produced in the United States and exported, whereas they account for 24.6 percent of total lifecycle emissions for gas produced and burned in Russia. This is because the gas produced in Russia must be transported for longer distances through pipelines, and methane leaks from these pipelines.[60]

With respect to the economic studies commissioned by the DOE for its public interest determination required for exports, the studies reached somewhat different conclusions. The Energy Information Administration study noted that the economic effects of export vary considerably depending on whether export amounts of natural gas are large or small, the speed at which these export amounts ramp up, and the amount of natural gas produced domestically. Looking across all of these scenarios, the study concluded that under all scenarios, domestic natural gas prices would increase, with the percentage changes in price ranging from 4 percent to 11 percent depending on the scenario.[61] Furthermore, consumers who use natural gas would pay higher natural gas bills on average between 2015 and 2040, with bills increasing by 1 percent to 8 percent over the baseline case; electricity bills would rise 0 percent or 3 percent, depending on the scenario.[62] However, some of these negative impacts would be balanced out by the fact that expanding LNG exports would "result in higher levels of economic output, as measured by real gross domestic product."[63] The study projected that more gas exports would cause GDP to rise 0.05 percent to 0.17 percent over the baseline scenario GDP, again depending on assumptions regarding the

---

[57] *Id.* at 11.
[58] *Id.*
[59] *Overview of Greenhouse Gases*, ENVTL. PROT. AGENCY, https://www3.epa.gov/climatechange/ghgemissions/gases/ch4.html.
[60] SKONE ET AL., *supra* note 54, at 11.
[61] U.S. ENERGY INFO. ADMIN., *supra* note 49, at 12.
[62] *Id.*
[63] *Id.*

volume of gas exported, the rapidity at which exports increased, and the rise in natural gas production in order to support exports.[64]

The NERA study on the economic impacts of natural gas exports from the United States focused somewhat more on the positive effects of export. This study noted that domestic prices of natural gas would only rise so much with increasing exports because foreign "importers will not purchase U.S. exports if the U.S. wellhead price rises above the cost of competing supplies."[65] In other words, if the cost of natural gas in the United States goes up dramatically as a result of reduced supply (due to higher exports of gas, which would cause less gas to be available domestically), other countries might have cheaper gas to export, and U.S. exports would drop. This, in turn, could cause domestic prices to decline as domestic supply increased. NERA also found that in all of the scenarios that it modeled, "the U.S. would experience net economic benefits from increased LNG exports," and that "U.S. economic welfare consistently increases as the volume of natural gas exports increase[s]."[66]

The DOE uses these studies in determining whether each proposed export of natural gas to a non-FTA country would be in the public interest. As of March 2016, it had received approximately 58 applications, most of which were for export to both FTA and non-FTA countries.[67] The agency had approved approximately 18 of the applications for export to non-FTA countries and nearly all of the applications to FTA countries, with the exception of a few applications that were vacated or were still pending approval.[68]

---

[64] Id.

[65] NERA ECONOMIC CONSULTING, *supra* note 50, at 6.

[66] Id.

[67] DEP'T OF ENERGY, LONG TERM APPLICATIONS RECEIVED BY DOE/FE TO EXPORT DOMESTICALLY PRODUCED LNG FROM THE LOWER–48 STATES (as of March 18, 2016), 1–3, http://energy.gov/sites/prod/files/2016/03/f30/Summary%20of%20LNG%20Export%20Applications.pdf.

[68] Id.

Chapter 8

THE SMART GRID AND DISTRIBUTED GENERATION

I. Introduction

The modern U.S electric utility industry began over 100 years ago with the growth of vertically integrated electric utility companies such as General Electric Company, Commonwealth Edison, and Westinghouse.[1] These companies and the many others that followed owned the electric generating facilities that produced electricity, the long-distance transmission lines that transported electricity from power plants to substations, and the distribution lines that delivered electricity to residential, commercial, and industrial customers. As discussed in detail in Chapter 6, the monopoly characteristics of these "investor-owned utilities" or IOUs led first to state regulation of electricity provision and price. Then in the early 20th century, as interstate electricity transactions grew—including sales across state lines—Congress enacted the Federal Power Act to regulate those interstate transactions and created the Federal Power Commission (now FERC), which, in turn, has promulgated numerous rules and orders governing wholesale electricity sales and prices and interstate electricity transmission. The 20th century also saw the rise of municipally owned utilities ("munis"); rural electric cooperatives, in which all customers are partial owners of a utility governed by an elected board; and power marketers. Together, these types of utilities serve just over 30 percent of electricity users in the United States.[2]

To take advantage of economies of scale, most IOUs and many municipal utilities and electric cooperatives historically relied on coal to power large, central station generating plants because coal was readily available, easy to store on site, and inexpensive. Hydropower was also heavily used in areas where it was available. The latter part of the 20th century saw a rise first in nuclear power plants and later

---

[1] Bernard W. Carlson, *Competition and Consolidation in the Electrical Manufacturing Industry, 1889–1892*, in TECHNOLOGICAL COMPETITIVENESS: CONTEMPORARY AND HISTORICAL PERSPECTIVES ON THE ELECTRICAL, ELECTRONICS, AND COMPUTER INDUSTRIES 289 (William Aspray ed., 1993).

[2] *See* AM. PUB. POWER ASS'N, 2015–2016 ANNUAL DIRECTORY & STATISTICAL REPORT 26. The percentage breakdown by number of customers is as follows: IOUs 68.4%, electric cooperatives 12.8%, power marketers (almost exclusively in Texas), 4.3%, and publicly owned or municipal utilities 14.5%. *Id.* The percentage breakdown by number of electricity providers is quite different: IOUs 5.7%, electric cooperatives 26.5%, federal power agencies 0.3%, and publicly owned utilities 60.9%. *Id.* Thus, there are a fairly small number of IOUs in the United States as compared to other types of electricity providers, but they serve the bulk of U.S. electricity customers.

in natural gas-fired power plants, which created alternative sources of electricity generation. Also in the latter part of the 20th century, as a result of federal statutes, FERC regulation, and state policies, non-utility generators began to compete with IOUs to provide electric energy in wholesale and markets and, in some states, retail markets as well.

With the dawn of the 21st century, developments in the "smart grid" (which is a grid that relies more on digital technologies to control grid functions) and distributed (i.e., localized) generation have created new opportunities to create more sophisticated and resilient electricity generation, transportation, and distribution systems. These developments include more efficient transmission and distribution of energy; two-way communications between electricity consumers and providers, in which consumers can more easily modify use in response to fluctuating electricity prices; new market opportunities for non-utility generators to provide electricity to customers; and the increasing ability for consumers themselves to generate and sell electricity. "Microgrids" are increasingly employed to allow colleges, hospitals, and communities to partially isolate the particular set of distribution wires and generation on which they rely and to be less dependent on the regional electric grid in times of bad weather and other disruptions, creating additional energy security options. But these developments have also created new threats to the IOU business model, requiring a new vision of the public utility as well as new federal and state regulations, policies, and markets. The rise of electric vehicles (EVs) plays a major role in this transition. EVs provide the opportunity to move away from petroleum as the dominant fuel in the transportation sector while also serving an integral role in smart grid development by providing distributed energy storage capabilities. (Consumers could offer their EVs, which store energy in batteries, as a back-up source of power for utilities to draw from during periods of peak demand, for example.) This Chapter discusses these issues in detail.

The first section of this Chapter explores the potential of the smart grid—what it does, the generation and grid improvements it can provide, and the growing attention to privacy issues associated with the increased collection and use of energy consumption data. The second section discusses the rise of EVs and the promise and challenges associated with relying more heavily on electricity to power the nation's transportation sector. Next, the third section focuses on the growth of energy storage technology and microgrids, both of which have the potential to reduce dependence on the traditional "macrogrid." With regard to energy storage, technology developments in battery storage and other storage mechanisms have the potential to enhance both microgrids and the macrogrid by eliminating the need to use electricity at the exact time it is generated, thus allowing significantly greater reliance on renewable energy resources such as wind and solar, which are "variable" or "intermittent." The output from these resources changes over time, and they can only be ramped up or down (decreasing or increasing electricity output, as needed) in limited ways.

II. The "Smart Grid"

The "Smart grid" describes a suite of technologies used to enhance electricity generation, transmission, and distribution systems "using digital technology that allows for two-way communication between the utility and its customers."[3] These systems are beginning to be used more extensively in all aspects of the electric grid—from central power plants and wind farms to long-distance transmission lines, home appliances, and other consumer electricity uses.[4] According to the U.S. DOE, the future smart grid would consist of:

> ... a fully automated power delivery network that monitors and controls every customer and node, insuring a two-way flow of electricity and information between the power plant and the appliance, and all points in between. Its distributed intelligence, coupled with broadband communications and automated control systems, enables real-time market transactions and seamless interfaces among people, buildings, industrial plants, generation facilities, and the electric network.[5]

The benefits associated with the smart grid include more efficient electricity transmission, reduced peak electric demand, increased integration of large scale and distributed renewable energy resources such as rooftop PV solar, more rapid restoration of electricity after power outages, heightened grid security, reduced operation and management costs for utilities, and lower electricity costs for consumers. Smart grid technologies include "smart meters"—two-way electricity meters that record electricity consumption in hourly or sub-hourly intervals and automatically convey that information back to the utility—as well as other technologies to enhance various aspects of the electricity generation, transmission, and distribution systems, such as synchrophasors,

---

[3] *What is the Smart Grid?*, SMARTGRID.GOV, https://www.smartgrid.gov/the_smart_grid/smart_grid.html.

[4] *Id.*

[5] U.S. DEP'T OF ENERGY, OFFICE OF ELECTRIC TRANSMISSION AND DISTRIBUTION, "GRID 2030" A NATIONAL VISION FOR ELECTRICITY'S SECOND 100 YEARS 17 (July 2003), energy.gov/sites/prod/files/oeprod/DocumentsandMedia/Electric_Vision_Document.pdf; RICHARD J. CAMPBELL, CONG. RESEARCH SERV., THE SMART GRID AND CYBERSECURITY: REGULATORY POLICY AND ISSUES (June 15, 2011).

which "record grid conditions with great accuracy and offer insight into grid stability or stress";[6] grid tie inverters, which meet "standards for safely connecting [distributed renewable resources] to or disconnecting [these resources] from the power grid based on monitored grid conditions" and ensure that the electricity produced by these resources is "compatible with the power grid in terms of voltage" and other conditions;[7] and automated substations.[8]

As of July 2014, utilities had installed more than 58 million smart meters, approximately 88 percent of which were for residential customers.[9] The enhanced two way communications smart meters enable real time pricing (pricing of electricity at different rates depending on the time at which it is delivered); facilitate deployment of distributed energy resources such as rooftop solar PV; increase energy efficiency; and enable "demand response"[10] programs that allow consumers to reduce their electricity use during periods of peak electricity demand and sell this electricity "non-use" into retail or wholesale markets.[11] Smart meter implementation is not uniform among either states or electric utilities. Some states like California, "Texas, and Arizona have installed smart meters for over 50 percent of customers, while others like Minnesota, New York, and Iowa are below 15 percent of customer meters."[12] The same is true for utilities, with some utilities installing millions of smart meters for their customers and others lagging behind with only hundreds installed.[13] In some jurisdictions, consumers have opposed smart meter

---

[6] *Synchrophasor Applications in Transmission Systems*, SMARTGRID.GOV, https://www.smartgrid.gov/recovery_act/program_impacts/applications_synchrophasor_technology.html.

[7] RICHARD MICHAEL, JR., PPL ELECTRIC UTILITIES, GRID-TIED SOLAR INVERTER REQUIREMENTS 25, https://www.pplelectric.com/~/media/pplelectric/at%20your%20service/docs/remsi/metering-equipment-tables/grid-tied-inverter-guide.pdf.

[8] *See* LINCOLN L. DAVIES ET AL., ENERGY LAW AND POLICY 710–12 (West Academic Publishing 2015) (providing an overview of smart grid technologies).

[9] *Frequently Asked Questions: How Many Smart Meters Are Installed in the United States, and Who Has Them?*, U.S. ENERGY INFO. ADMIN., May 16, 2014.

[10] "Demand response provides an opportunity for consumers to play a significant role in the operation of the electric grid by reducing or shifting their electricity usage during peak periods in response to time-based rates or other forms of financial incentives. Demand response programs are being used by electric system planners and operators as resource options for balancing supply and demand." *Demand Response*, U.S. DEP'T OF ENERGY, http://energy.gov/oe/technology-development/smart-grid/demand-response. For further discussion of demand response and court decisions regarding FERC's regulation of demand response pricing, see *infra* notes 30–31 and accompanying text.

[11] *See id.*

[12] Alexandra B. Klass & Elizabeth J. Wilson, *Energy Consumption Data: The Key to Improved Energy Efficiency*, 6 SAN DIEGO J. CLIMATE & ENERGY L. 69, 75 (2014–2015); *Smart Meter Deployments Continue to Rise*, U.S. ENERGY INFO. ADMIN. (Nov. 1, 2012), http://www.eia.gov/todayinenergy/detail.cfm?id=8590.

[13] *See* INST. FOR ELEC. INNOVATION, UTILITY-SCALE SMART METER DEPLOYMENTS: BUILDING BLOCK OF THE EVOLVING POWER GRID (Sept. 2014).

implementation because of fears that they will compromise human health, safety, and privacy.[14] With regard to public health, numerous state departments of health and the American Cancer Society have concluded that the health risks associated with smart meters are minimal, but citizen concerns remain.[15]

Policies and incentives to enhance the smart grid exist at the federal, state, and local levels. In the Energy Policy Act of 2005 (EPAct 2005),[16] Congress made it "the policy of the United States to encourage . . . State energy policies to provide reliable and affordable demand response services to the public"[17] and required DOE to work "with States, utilities, other energy providers and advanced metering and communications experts to identify and address barriers to the adoption of demand response programs" and to identify national benefits of these programs.[18] Congress additionally directed FERC to "prepare and publish an annual report . . . that [inter alia] assesses demand response resources" and that identifies barriers to "customer participation" in these programs, including the ability of customers to reduce energy use during "critical" (peak) periods of electricity use.[19] As introduced above, these programs encourage industrial, commercial, and residential utility customers to reduce electricity load during peak periods so that there is less "congestion" in transmission and distribution lines and utilities do not have to rely as much on more expensive generating plants that only run during peak hours. The smart grid makes demand response participation easier by providing up-to-date information on when reduced electricity demand is needed. EPAct 2005 also required state PUCs and other entities regulating electricity providers to determine whether utility customers in their jurisdiction should be required to offer customers time-based rates,[20] which vary based on customer demand during different parts of the day, and smart meters.[21] The Act further required utilities to provide this type of pricing if customers requested it.[22]

---

[14] *See* Stop Smart Meters!, http://stopsmartmeters.org/; Andy Balaskovitz, *Despite Court Setbacks, Michigan Smart-Meter Opponents 'Not Going Away,'* MIDWEST ENERGY NEWS, July 28, 2015 (reporting on series of three decisions by Michigan Court of Appeals upholding efforts by Michigan's two largest utilities to install smart meters in customer homes and rejecting efforts by smart meter opponents to prevent installation of smart meters on privacy and public health grounds).

[15] *See* DAVIES ET AL., *supra* note 8.

[16] Pub. L. No. 109–58, 119 Stat. 594 (2005).

[17] *Id.* at § 1252(e), codified at 16 U.S.C. § 2642.

[18] *Id.* at § 1252(d), codified at 16 U.S.C. § 2642.

[19] *Id.* at § 1252(e), codified at 16 U.C.C. § 2642.

[20] *Id.* at § 1252(b), codified at 16 U.S.C. § 2625.

[21] *Id.*

[22] *Id.*

With the Energy Independence and Security Act of 2007 (EISA 2007),[23] Congress went further and included a "Statement of Policy on Modernization of the Electric Grid" which declared it the policy of the United States to "support the modernization of the Nation's electricity transmission and distribution system to maintain a reliable and secure electricity infrastructure that can meet future demand growth . . . "[24] This policy statement included numerous goals for "modernizing" the transmission and distribution grid to support the following types of efforts: better integrating renewable resources and distributed energy resources such as rooftop PV solar; integrating "smart" appliances that can communicate with utilities through smart meters and, for truly "smart" appliances, can be automatically cycled up and down by a remote utility computer; and deploying and integrating energy storage.[25] EISA 2007 also required the creation of smart grid taskforces at the federal and regional levels. [26]

Another critical component of EISA 2007 was language that directed the National Institute of Standards and Technology (NIST) to work with DOE, FERC, and others to develop a Smart Grid Interoperability Framework.[27] This Framework was intended to make it easier for a variety of customers to seamlessly connect to and communicate with the grid in a uniform, two-way fashion, allowing them to plug in any appliance, electric car, or other electricity-consuming or electricity-producing apparatus (such as a solar PV panel) and have that apparatus immediately communicate with the grid and receive information from the grid, such as information on the current price of electricity. For example, Congress hoped that, through interoperability standards, electric car owners could plug in their cars at night and program their cars to automatically start charging a battery when prices dropped below a particular level, or even to allow the utility to draw electricity from the battery when demand went up and electricity prices spiked. Ideally, all appliances in a building would even be able to communicate with each other, and customers could program buildings as an entire "package," with certain parts of buildings consuming more or less electricity during different periods depending on how essential those technologies were and what the price of electricity was. Further, interoperability standards were intended to allow all other devices on the grid—such as computers that controlled transmission and distribution lines, to "talk" to each other in the event that there was a problem in one line.

---

[23] Pub. L. No. 110–140, 121 Stat. 1492 (2007).
[24] *Id.* at § 1301, codified at 42 U.S.C. § 17381.
[25] *Id.*
[26] *Id.* at § 1303(b), codified at 42 U.S.C. § 17383.
[27] *Id.* at § 1305, codified at 42 U.S.C. § 17385.

In short, the goal was to make the grid similar to the Internet, allowing devices to easily connect with and communicate with each other. Many of the directives of EISA 2007 are well underway.

Two years later, in 2009, the American Recovery and Reinvestment Act provided $4.5 billion in funding for smart grid technologies[28] and included significant resources for new smart meters nationwide through grants to states and utilities.[29] This funding has helped to support certain EISA goals.

FERC and state PUCs have also been active in smart grid developments. FERC issued Order 745 in 2011, which provided incentive pricing for demand response.[30] Specifically, this Order required that entities offering reduced electricity use into wholesale energy markets in lieu of generation—allowing grid operators to rely on electricity non-use during peak periods rather than having to draw from more expensive generation—should receive the same price for their product that other generators received for providing peak wholesale electricity at that that time in that particular region, as long as the demand response resource was in fact cheaper than alternatives.[31] The U.S. Supreme Court affirmed FERC's authority to provide such incentives in 2016 in *FERC v. Electric Power Supply Association*.[32] FERC and state PUCs have also addressed grid functionality, integrating variable renewable energy resources, and developing and incentivizing EV fleets and charging infrastructure.

Challenges facing the development of the smart grid include finding the right incentives to encourage utilities to innovate, informing consumers about the benefits of the smart grid, encouraging state PUCs to allow utilities to recover the costs associated with implementing smart grid technologies, and encouraging both state PUCs and utilities to create and incentivize demand response.[33] Another challenge associated with the smart grid is the need to collect and analyze "energy consumption data (also

---

[28] Klass & Wilson, *supra* note 12, at 82.

[29] *Recovery Act Smart Grid Programs*, U.S. DEP'T OF ENERGY, OFFICE OF ELEC. DELIVERY & ENERGY RELIABILITY, https://www.smartgrid.gov/recovery_act.

[30] Demand Response Compensation in Organized Wholesale Energy Markets, 76 Fed. Reg. 16,658, (Mar. 24, 2011) (codified at 18 C.F.R. pt. 35).

[31] *Id.*

[32] Fed. Energy Regulatory Comm'n v. Elec. Power Supply Ass'n, 136 S. Ct. 760 (2016). In its decision, the Court held that FERC's authority to regulate wholesale electricity sales under the Federal Power Act extended to the regulation of demand response. The Court recognized that demand response programs eased pressure on the electric grid, reduced the need for new electricity generation, and promoted lower wholesale electricity prices.

[33] *See, e.g.*, Joel B. Eisen, *Smart Regulation and Federalism for the Smart Grid*, 37 HARV. ENVTL. L. REV. 1 (2013).

referred to as 'customer energy usage data')"[34] and address the privacy concerns that often accompany the increased availability of this data. Despite the extensive investments in the smart grid detailed above, a major impediment to making full use of the smart grid is the lack of readily available energy consumption data. With better energy consumption data, utilities would have greater awareness of the times of day and areas where more electricity would be needed and the opportunities for drawing from customer-provided demand response resources during periods of peak demand, among other opportunities. Likewise, customers and third party energy efficiency providers could use that data to significantly increase energy efficiency in homes, businesses, and industries.

Even though utilities collect energy consumption data for billing purposes, the data is "surprisingly difficult for governments, energy efficiency service providers, and researchers to obtain and evaluate."[35] As explained in previous work:

> In the past, when electricity meter reading was not automated and was conducted only on a monthly basis, the difficulty in accessing and evaluating such data was understandable. But in today's big-data world of intra-hourly electricity use recorded by smart meters, transmitted to utilities wirelessly or through fiber networks, and with electronic billing to customers and new technologies emerging for energy management and generation, energy consumption data has become a critical linchpin in the electric sector and is used regularly by electric utilities in their daily operations.[36]

Nevertheless, the data is not often collected and organized in a standardized format by utilities. This makes it difficult to fulfill many of the goals of the smart grid—increased energy efficiency in buildings, increased implementation of distributed resources, implementation of demand response programs, reduction of GHG emissions, and lower electricity prices.[37] Even when standardized

---

[34] Klass & Wilson, *supra* note 12, at 73. Energy consumption data is a record of the amount of electricity a building uses over a given period of time. *See, e.g.*, *Best Practices for Working with Utilities to Improve Access to Energy Usage Data*, AM. COUNCIL FOR AN ENERGY-EFFICIENT ECON. (June 2014), http://aceee.org/sector/local-policy/toolkit/utility-data-access. Energy consumption data can also include natural gas and water use in homes and businesses.

[35] Alexandra B. Klass & Elizabeth J. Wilson, *Remaking Energy: The Critical Role of Energy Consumption Data*, CAL. L. REV. (forthcoming 2016), *available at* http://papers.ssrn.com/sol3/papers.cfm?abstract_id=2659343.

[36] *Id.*

[37] *See, e.g.*, ABRAMS ENVIRONMENTAL LAW CLINIC, UNIVERSITY OF CHICAGO LAW SCHOOL, FREEING ENERGY DATA: A GUIDE FOR REGULATORS TO REDUCE ONE BARRIER TO RESIDENTIAL ENERGY EFFICIENCY 4-8 (June 2016).

data is available, some consumers and privacy groups express concern over making such data available to third parties without customer consent.

With regard to the issue of data standardization, in 2011, the White House issued a challenge for electricity providers to make energy consumption data more readily available to customers in a uniform format, and the energy sector responded by developing the "Green Button" initiative.[38] Since its official launch in 2012, "more than thirty-five utilities and electricity suppliers have adopted" Green Button.[39] As explained in previous work:

> The standard means that utilities can follow a uniform approach to data collection and presentation, allowing EESPs to develop software more easily to analyze the data and recommend efficiency improvements to consumers, rather than develop software specific to each utility's data set. Green Button data can be provided in 15-minute, hourly, daily, or monthly intervals depending on what the utility decides to make available and the level of detail it is able to provide.[40]

Nevertheless, the program is voluntary and, to date, only a limited number of electric utilities have adopted it.[41]

As for privacy concerns, federal and state regulators and consumers have expressed fears that the release of energy consumption data, particularly sub-hourly data tied to a particular address, could fall into the wrong hands and be used to assist in criminal activity. For instance, a "burglar could determine times of day a residence is likely unoccupied."[42] "Further, [t]he data could be

---

[38] Klass & Wilson, *supra* note 12, at 84–85; *see also Green Button*, ENERGY.GOV, http://www.energy.gov/data/green-button; *Green Button*, PAC. GAS & ELEC., http://www.pge.com/myhome/addservices/moreservices/greenbutton/; Nick Sinai & Matt Theall, *Expanded "Green Button" Will Reach Federal Agencies and More American Energy Consumers*, THE WHITE HOUSE, Dec. 5, 2013, http://www.whitehouse.gov/blog/2013/12/05/expanded-green-button-will-reach-federal-agencies-and-more-american-energy-consumers.

[39] Klass & Wilson, *supra* note 12, at 85; *see also* Monisha Shah & Nick Sinai, *Green Button: Enabling Energy Innovation*, WHITE HOUSE BLOG, May 2, 2013, https://www.whitehouse.gov/blog/2013/05/02/green-button-enabling-energy-innovation. For a list of the total entities, now numbering approximately seventy-five, which have adopted the program, see GREEN BUTTON DATA.ORG.

[40] Klass & Wilson, *supra* note 12, at 85.

[41] Resistance to the program is strongest from utilities that have developed their own proprietary data-sharing software that is not consistent with the Green Button standards. *See, e.g.*, In the Matter of the Proposed Rules Relating to Data Access and Privacy for Electric Utilities, 4 Code of Colorado Regulations 723-3 and Data Access and Privacy Rules for Gas Utilities, 4 Code of Colorado Regulations 723-4, 2015 WL 2089032 ¶ 132 (Colo. P.U.C., May 1, 2015), adopted as modified, 2015 WL 4572600 (Colo. P.U.C. July 23, 2015).

[42] Klass & Wilson, *supra* note 35, at 18.

used for marketing purposes, or could expose criminal activity or zoning violations."[43] The states have had to respond to these concerns; "neither Congress nor any federal agency has created specific privacy policies governing energy consumption data."[44]

Some efforts to improve consumer data privacy have begun:

> [In] 2015, the DOE released a Voluntary Code of Conduct ("VCC") on data privacy and the smart grid. The VCC was intended to instill consumer confidence by addressing privacy concerns regarding energy consumption data. It specifies policies for: Customer Notice and Awareness, Customer Choice and Consent, Customer Data Access, Data Integrity and Security, and Self-Enforcement Management and Redress. . . . [One section of the VCC] details[s] the release of data [and] does not require individual consent. This release is permissible if "the methodology used to aggregate or anonymize Customer Data strongly limits the likelihood of re-identification of individual customers or the Customer Data from the aggregated or Anonymized data set." The VCC defines "aggregated data" as "a combination of data elements for multiple customers to create a data set that is sufficiently anonymous so that it does not reveal the identity of an individual customer."[45]

At the state level, legislatures and PUCs "have taken a variety of approaches to make energy consumption data available to customers and third parties,"[46] and regulatory proceedings are underway in numerous states.[47] Although not all states have taken steps to improve data availability, those that have considered the issue, such as California, Colorado, Illinois, Oklahoma, Pennsylvania, Texas, and Washington, have opened up access to information, at least for customers.[48] Other states have gone farther—"[s]ome, including Washington, require data to be provided in specific formats, including those compatible with EPA's ENERGY STAR Portfolio Manager or Green Button."[49] As for the ability of cities, researchers, energy efficiency providers or other third parties

---

[43] *Id.*

[44] *Id.*

[45] Klass & Wilson, *supra* note 35, at 16 (quoting U.S. DEP'T OF ENERGY, VOLUNTARY CODE OF CONDUCT (VCC): FINAL CONCEPTS AND PRINCIPLES (Jan. 12, 2015)).

[46] *Id.* at 17.

[47] For a fuller discussion of state and local policies regarding energy consumption data and privacy, see Klass & Wilson, *supra* note 35; Klass & Wilson, *supra* note 12.

[48] Klass & Wilson, *supra* note 35, at 18.

[49] *Id.*

to access energy consumption data, states have begun to create a variety of rules regarding data aggregation and anonymization to make consumer data more readily available.[50]

This section has shown that the Smart Grid has a number of functions, from making the flow of electricity through transmission and distribution lines more efficient to allowing customers to better control when and how they use electricity. This latter function is a form of energy efficiency; customers with better information about their electricity consumption and the appliances that use the most electricity can use technology like "smart" appliances to turn these appliances off, or cycle down their usage (in the case of air-conditioning, for example), when they are not needed or when electricity prices are high. There are also more basic approaches to energy efficiency that do not require sophisticated technologies, such as weatherizing buildings by installing more caulking around windows and doors or installing energy-efficient windows and light bulbs.

Numerous federal and state initiatives have encouraged improved energy efficiency along with smart grid improvements. For example, many states require "integrated resource planning" (IRP)— a process by which utilities plan for how they will provide electricity to meet future demand and report this plan to the states. Through the IRP process, states can require utilities to consider how, instead of building new generation to meet demand, they will decrease demand through energy efficiency programs.[51] Other states have "Energy Efficiency Resource Standards," which require utilities or other electricity-providing entities to reduce a certain percentage of electricity use by improving the energy efficiency of their customers.[52] As of March 2015, approximately 18 states had these types of standards, and additional states had energy efficiency goals.[53] Further, states and local governments provide extensive funding for weatherizing homes and improving the energy efficiency of buildings in other ways, often with the help of federal funds.[54]

---

[50] *Id.* at 17–19.

[51] *See, e.g.,* RACHEL WILSON & BRUCE BIEWALD, SYNAPSE ENERGY ECON., BEST PRACTICES IN ELECTRIC UTILITY INTEGRATED RESOURCE PLANNING: EXAMPLES OF STATE REGULATIONS AND RECENT UTILITY PLANS 7 (June 2013) (prepared for the Regulatory Assistance Project), http://www.synapse-energy.com/sites/default/files/SynapseReport.2013-06.RAP_.Best-Practices-in-IRP.13-038.pdf (describing some states' requirements for integrated resource plans).

[52] *Energy Efficiency Resource Standards (EERS),* AM. COUNCIL FOR AN ENERGY-EFFICIENT ECON., http://aceee.org/topics/energy-efficiency-resource-standard-eers.

[53] Database of State Incentives for Renewables & Efficiency, Energy Efficiency Resource Standards (and Goals) (March 2015), http://ncsolarcen-prod.s3.amazonaws.com/wp-content/uploads/2015/03/Energy-Efficiency-Resource-Standards.pdf.

[54] U.S. DEP'T OF AG., ET AL., FEDERAL FINANCE FACILITIES AVAILABLE FOR ENERGY EFFICIENCY UPGRADES AND CLEAN ENERGY DEPLOYMENT (Aug. 28, 2013).

The federal government—in the same acts that required efforts to expand the smart grid—also included certain energy efficiency mandates, grants, and tax credits. For example, in the Energy Policy Act of 2005 Congress required federal agencies to include energy efficient products in their procurement plans[55] and directed the Secretary of Energy to "establish an Advanced Building Efficiency Testbed program for the development, testing, and demonstration of advanced engineering systems, components, and materials to enable innovations in building technologies."[56] Further, Congress required the Secretary of Energy to write more stringent "Federal building energy efficiency standards";[57] provided funds for state energy efficient appliance rebates[58] and for grants to states that help local governments improve "the energy efficiency of public buildings and facilities";[59] and provided for grants to states, Native American tribes, and certain organizations for energy efficiency in low-income communities, among other provisions.[60] In EISA 2007 Congress required newly-manufactured motors, air conditioners, lamps and other equipment to meet certain efficiency standards;[61] provided cost-sharing for the development of "zero-net-energy commercial buildings"—those that "result in" no net emissions of greenhouse gases";[62] and required more energy efficiency efforts for federal buildings,[63] among other energy efficiency initiatives. Thus, while the smart grid and improved measurement of energy use are key tools for improving energy efficiency, they are parts of a much broader movement to reduce energy use.

III. Electric Vehicles (EVs)

One component of the smart grid is the potential to integrate large numbers of EVs into the grid, allowing those vehicles to charge at night when electricity costs are low and serve as a form of energy storage when electricity supply exceeds demand. In recent years, an increasing number of federal and state policies have been implemented to encourage the development of alternative vehicles to reduce the use of gasoline in the transportation sector. Reasons for wanting to reduce U.S. reliance on petroleum include energy security, fuel costs, the carbon emissions and other pollution associated with extracting and refining oil, and the tailpipe

---

[55] Pub. L. no. 109–58 § 104 (Aug. 8, 2005), codified at 42 U.S.C. § 8259b.
[56] Id. at § 107, codified at 42 U.S.C. § 15812.
[57] Id. at § 109, codified at 42 U.S.C. § 6834.
[58] Id. at § 124, codified at 42 U.S.C. § 15821.
[59] Id. at § 125, codified at 42 U.S.C. § 15822.
[60] Id. at § 126, codified at 42 U.S.C. § 15823.
[61] Pub. L. no. 110–140 § 314 (Dec. 19, 2007), codified at 42 U.S.C. §§ 6311, 6295.
[62] Id. at § 422, codified at 42 U.S.C. § 17082.
[63] Id. at § 432, codified at 42 U.S.C. § 8253.

emissions associated with combusting petroleum products in internal combustion engine vehicles (ICEVs), which cause significant air pollution in urban areas. Alternative vehicles include hybrid electric vehicles (HEVs), such as the Toyota Prius, which a include a fuel tank and a battery that captures energy from braking; partial hybrid electric vehicles (PHEVs), such as the Chevy Volt, which include a battery charged by an electric cord (and which also captures energy from braking), and a fuel tank; and all-electric vehicles (EVs), such as the Tesla and the Nissan Leaf. All of these alternative vehicles offer a much higher rate of fuel efficiency than traditional internal combustion engine vehicles (ICEVs). Nevertheless, fluctuating gasoline prices, consumer wariness of new technology, and the difficulty of creating the electric charging station infrastructure needed for the use of EVs has made widespread adoption of all of these technologies slower than many desire.[64] The EIA estimates that by 2040, alternative vehicles of all types will make up approximately 13 percent of the nation's automotive stock.[65]

Among the states, California has always been a national leader in setting stringent vehicle emissions standards, and was the first state to incentivize the production, marketing, and use of EVs by creating a Zero-Emission Vehicle (ZEV) program in 1990.[66] In recent years, the California Air Resources Board (CARB) has substantially strengthened the program, with current regulations requiring 22 percent of cars of model years 2025 and later to be ZEVs.[67] California also worked with the EPA and NHTSA to develop what was to become the 2012 federal rule limiting GHGs from light-duty vehicles.[68] Other states have followed California's lead and, as of 2015, approximately one third of the nation's new car sales occurred in a state governed by the ZEV rule.[69]

Beyond the mandate in some states that car manufacturers and dealers offer a certain number of EVs for sale, there are federal and state tax incentives to encourage the purchase of EVs. At the federal level there is a federal tax credit of up to $7,500 for the purchase of

---

[64] U.S. EIA estimates that in 2015, there were 3.09 million HEVs in use and 100,000 EVs in use. U.S. Energy Info. Admin., Annual Energy Outlook 2015, Table 40, http://www.eia.gov/beta/aeo/#/?id=49-AEO2015.

[65] *Id.* EIA includes as "alternative vehicle" for this prediction gasoline-electric hybrids and ethanol-fueled internal combustion engines, which together make up over 75% of the estimated total. *Id.*

[66] Louise Wells Bedsworth & Margaret R. Taylor, *Learning from California's Zero-Emission Vehicle Program*, 3 CAL. ECON. POLICY, no. 4 (Sept. 2007).

[67] 13 CCR § 1962.2 (West 2016).

[68] *ARB Proposes Regulations to Accept Federal GHG Vehicle Standards*, CAL. ENVTL. PROT. AGENCY, AIR RESOURCES BD., Aug. 30, 2012, http://www.arb.ca.gov/newsrel/newsrelease.php?id=348.

[69] *Federal Tax Credits for All-Electric and Plug-in Hybrid Vehicles*, U.S. DEP'T OF ENERGY, https://www.fueleconomy.gov/feg/taxevb.shtml.

an EV or PHEV.[70] Many states have additional tax incentives ranging from $1,000 to $6,000 to encourage the purchase of EVs for private[71] and corporate[72] use, and some states have mandated state agency purchase of EVs or other lower emissions vehicles.[73] These subsidies increase demand for EVs by lowering their net costs, while state ZEV mandates create a market for EVs and encourage automakers to participate in it.

Several challenges are associated with the widespread adoption and use of EVs. First, current limits on EV battery technology, coupled with the lack of fast, readily available vehicle charging stations, leads to "range anxiety," which reduces the demand for EVs. At the present time, more expensive EVs such as the Tesla can travel up to 215 miles on a charge,[74] but more modestly priced EVs such as the Nissan Leaf can travel 107 miles in warm weather.[75] The Leaf's range is shorter in very cold weather,[76] which is a problem for EV adoption in northern climates. Charging stations are not nearly as readily available as gas stations and, in many parts of the United States, non-existent. Charging often takes several hours and, without a network of charging stations, EVs are often more suitable to urban driving and not practical for long-distance trips or even for daily use in rural areas. Technology developments and increasing incentives for a nationwide network of charging stations will continue to improve the market for EVs, but without a massive investment in infrastructure, progress will remain slow.

Second, although it may not be immediately obvious, the environmental benefits of EVs over ICEVs are not the same in all parts of the country. With regard to tailpipe emissions, EVs and PEVs (when running on the battery rather than on the hybrid gas engine) have no tailpipe emissions at all, so they will always be environmentally superior to ICEVs if those are the only emissions that are considered in the comparison. But once the focus expands

---

[70] *Federal Tax Credits for All-Electric and Plug-In Hybrid Vehicles*, U.S. DEP'T OF ENERGY, https://www.fueleconomy.gov/feg/taxevb.shtml; LYNN J. CUNNINGHAM ET AL., CONG. RES. SERV., ALTERNATIVE FUEL AND ADVANCED VEHICLE TECHNOLOGY INCENTIVES: A SUMMARY OF FEDERAL PROGRAMS 7–8 (Jan. 10, 2013).

[71] *See, e.g.*, COLO. REV. STAT. § 39–26–719 (West 2016).

[72] *See, e.g., Illinois Green Fleets Program*, ILLINOIS ENVTL. PROT. AGENCY, http://www.illinoisgreenfleets.org/.

[73] *See, e.g.*, MINN. STAT. §§ 16C.137–.138 (West 2016); *State Efforts Promote Hybrid and Electric Vehicles*, NAT'L CONFERENCE OF STATE LEGISLATURES (Dec. 3, 2015).

[74] Model 3, Tesla, https://www.teslamotors.com/model3.

[75] 2016 Nissan Leaf®, Nissan, http://www.nissanusa.com/electric-cars/leaf/.

[76] Matthew Klippenstein, *Nissan Leaf, Chevy Volt Range Loss in Winter: New Data from Canada*, GREEN CAR REPORTS, Dec. 17, 2013, http://www.greencarreports.com/news/1089160_nissan-leaf-chevy-volt-range-loss-in-winter-new-data-from-canada.

beyond tailpipe emissions to include the emissions associated with the production and transportation of the energy resources necessary to produce the electricity to power EVs, the benefits of EVs in some parts of the United States are more questionable.[77] Numerous experts have undertaken "life cycle analysis"[78] studies to address this issue. Life cycle analysis investigates the environmental impacts of all phases of a product's life cycle, including extracting and transporting the minerals or other resources necessary for that product to manufacturing the product, using the product, and eventually disposing of it.[79] These life cycle studies focus primarily on the "use phase" of the vehicles—i.e., the emissions associated the generation, transportation, and combustion of energy resources used to power the vehicles. This is because "full life cycle" studies have mostly concluded that the emissions associated with production and disposal of the vehicles themselves (including batteries for the EVs) constitute only a small portion of the lifecycle emissions.

With regard to the use phase studies for EVs, the studies consider: (1) the emissions associated with the extraction and transportation of coal, natural gas, nuclear fuel, wind, hydropower, and solar energy used to produce electricity and (2) the emissions associated with converting those energy resources into the electricity needed to charge the EV, such as the emissions from a coal-fired power plant or natural gas fired plant.[80] For ICEVs, a use phase analysis considers the environmental impacts of producing oil and biofuels, transporting those fuels to refineries and blending facilities, refining and blending those fuels to turn them into gasoline and other vehicle fuels, and the tailpipe emissions from the combustion of these fuels to drive the ICEV.

Once these environmental impacts are considered, the environmental benefits of EVs as compared to ICEVs are mixed as a

---

[77] Eric Jaffe, *4 Key Problems with Measuring EV Pollution vs. Gas Cars*, THE ATLANTIC'S CITYLAB, July 10, 2015; Virginia D. McConnell & Joshua Linn, *Subsidies for EVs: Good Policy or Unnecessary Handout?*, RES. FOR THE FUTURE, July 25, 2013.

[78] *Defining Life Cycle Assessment*, GLOBAL DEV. RESEARCH CTR., http://www.gdrc.org/uem/lca/lca-define.html. In its early days life cycle analysis was primarily used for product comparisons such as comparing the environmental impacts of disposable and reusable products. Today its applications include government policy, strategic planning, marketing, consumer education, process improvement and product design. *Id.*

[79] SCIENTIFIC APPLICATIONS INTL. CORP. LIFE CYCLE ASSESSMENT: PRINCIPLES AND PRACTICE 1 (2006) (prepared for EPA), http://nepis.epa.gov/Exe/ZyPDF.cgi/P1000L86.PDF?Dockey=P1000L86.PDF.

[80] *See, e.g.*, KIMBERLY AGUIRRE ET AL., LIFECYCLE ANALYSIS COMPARISON OF A BATTERY ELECTRIC VEHICLE AND A CONVENTIONAL GASOLINE VEHICLE (June 2012) (prepared for Cal. Air Res. Bd.), http://www.environment.ucla.edu/media/files/BatteryElectricVehicleLCA2012-rh-ptd.pdf; T.R. Hawkins et al., *Comparative Environmental Life Cycle Assessment of Conventional and Electric Vehicles*, 17 J. IND. ECOL. 53 (2013).

result of the current state of the U.S. electric grid.[81] For instance, coal-fired electricity, which constitutes approximately one-third of electricity generation nationwide[82] but much higher and lower percentages in certain states, emits substantial amounts of criteria pollutants such as nitrogen oxides, sulfur dioxides, and particulate matter; toxic emissions such as mercury; and GHG emissions that contribute to climate change.[83] Thus, a major question with regard to EVs is whether the decrease in tailpipe emissions compared to ICEVs can overcome the electricity generation-related emissions and other use phase emissions.[84]

Taking the average mix of generation sources for U.S. electricity, average annual GHG emissions are lower for EVs, HEVs and PHEVs.[85] However, the wide variety of energy resources used to produce electricity among the various states means that focusing on national averages is not always useful.[86] For instance, based on 2014 data, West Virginia and Kentucky rely on coal for over 90 percent of state electricity generation, while other states like Missouri and North Dakota use coal for over 70 percent of state electricity generation.[87] Thus, plugging in and driving an EV, HEV, or PHEV vehicle in these states does not result in significant GHG emission reductions. By contrast, California and Maine do not use coal to

---

[81] *See, e.g., Cleaner than What?*, THE ECONOMIST, Dec. 20, 2014; Christopher W. Tessum, Jason D. Hill & Julian D. Marshall, *Life Cycle Air Quality Impacts of Conventional and Alternative Light-Duty Transportation in the United States*, 111 PROCEEDINGS OF THE NAT. ACAD. OF SCI. 18490, 18490 (2014); Alexandra B. Klass & Andrew Heiring, *Life Cycle Analysis and Transportation Energy*, BROOKLYN L. REV. (forthcoming 2016) (discussing life cycle studies for EVs).

[82] *What is U.S. Electricity Generation by Energy Source?*, U.S. ENERGY INFO. ADMIN., https://www.eia.gov/tools/faqs/faq.cfm?id=427&t=3.

[83] *See* Nicholas Z. Muller et al., *Environmental Accounting for Pollution in the United States Economy*, 101 AM. ECON. REV. 1649, 1664 (2011); *How Much of U.S. Carbon Dioxide Emissions are Associated with Electricity Generation?*, U.S. ENERGY INFO. ADMIN., http://www.eia.gov/tools/faqs/faq.cfm?id=77&t=11.

[84] To better enable comparisons between ICEVs and EVs/HEVs, the U.S. Department of Energy has developed an online calculator which estimates GHG emissions based on car model, make, and zip code. *Beyond Tailpipe Emissions—Fuel Economy*, U.S. DEP'T OF ENERGY, https://www.fueleconomy.gov/feg/Find.do?action=bt2.

[85] *Emissions from Hybrid and Plug-In Electric Vehicles*, U.S. DEP'T OF ENERGY, http://www.afdc.energy.gov/vehicles/electric_emissions.php. This study exclusively examined GHG emissions (carbon dioxide, methane, and nitrous oxide). A. Elgowainy, A. Burnham, M. Wang, J. Molburg & A Rousseau, *Well-to-Wheels Energy Use and Greenhouse Gas Emissions Analysis of Plug-in Hybrid Electric Vehicles*, CTR. FOR TRANSP. RESEARCH, ARGONNE NAT'L LAB. 25 (Feb. 2009).

[86] Considering the national average generation mix is more relevant in international comparisons. *See* Lindsay Wilson, *Shades of Green: Electric Cars' Carbon Emissions Around the Globe*, SHRINK THAT FOOTPRINT 25 (Feb. 2013).

[87] Alyson Hurt, *Coal, Gas, Nuclear, Hydro? How Your State Generates Power*, NAT'L PUB. RADIO, Sept. 10, 2015, http://www.npr.org/2015/09/10/319535020/coal-gas-nuclear-hydro-how-your-state-generates-power (using EIA data to show change in energy resources used to generate electricity in each of the 50 states from 2004–2014).

generate any in-state electricity.⁸⁸ States like California, Delaware, Florida, Mississippi, Nevada, Rhode Island, and Massachusetts rely on natural gas to generate a significant percentage of electricity in the state.⁸⁹ And Idaho, Washington, and Oregon rely mostly on hydropower for their electricity needs.⁹⁰ Despite the varied benefits of EVs, HEVs, and PHEVs among states, there are good arguments to continue to promote EVs nationwide rather than only in states with "cleaner" electric grids. In many traditionally coal-based states, like Minnesota, Iowa, and Kansas, the growth of renewable energy resources, particularly wind energy, and the low cost of natural gas means that coal is rapidly losing its dominance in the electricity sector, even in states where it has been prominent for decades.⁹¹ Moreover, nationwide, the use of coal to generate electricity has dropped from just over 50 percent to 40 percent from 2004–2014, and by the end of 2016, natural gas is expected to become the largest energy source for U.S. electricity.⁹² Indeed, in April 2015 (although not in other months in 2015), natural gas surpassed coal in fueling U.S. electricity generation.⁹³

IV. Energy Storage

Utilities and other companies are increasingly developing large-scale batteries to complement the current use of energy storage technologies.⁹⁴ As explained in previous work, "[e]nergy storage consists of a suite of technologies including batteries, pumped-storage hydropower, compressed air storage, flywheels, and thermal energy that retain energy from electricity generated at times of low demand, strong winds, or peak sun until demand increases."⁹⁵ Energy storage serves many purposes. It enhances the reliability of the grid by providing backup electricity during blackouts and

---

⁸⁸ Hurt, *supra* note 87.

⁸⁹ *Id.*

⁹⁰ *Id.*

⁹¹ *Id.; see also* Daniel Cusick, *Okla., Kan. Wind Helped Drive Record CO_2 Reductions-Study*, CLIMATEWIRE, Mar. 29, 2016.

⁹² Alexandra B. Klass, *Future-Proofing Energy Transport Law*, 94 WASH. U. L. REV. (forthcoming 2017), *available at* http://papers.ssrn.com/sol3/papers.cfm?abstract_id=2748905.

⁹³ *Electricity from Natural Gas Surpasses Coal for First Time, But Just for One Month*, U.S. ENERGY INFO. ADMIN., July 31, 2015, http://www.eia.gov/todayinenergy/detail.cfm?id=22312.

⁹⁴ Klass, *supra* note 92, at 17.

⁹⁵ *Id.* (citing Amy L. Stein, *Reconsidering Regulatory Uncertainty: Making a Case for Energy Storage*, 41 FLA. STATE U. L. REV. 697, 705–09 (2014)). The most common form of energy storage is pumped-storage hydropower, which uses electricity generated during off-peak hours (and thus is less expensive) to pump water from a lower to an upper reservoir and then releases the water to turn turbines to generate electricity during on-peak hours. There are approximately 22 gigawatts of pumped-storage hydropower in the United States at 40 different sites, most of which was built between 1970 and 1990. *Id.* at 705.

brownouts.[96] It helps grid operators maintain appropriate frequency regulation (the balance of electricity within the transmission and distribution grid, which must be maintained at a constant level) by adding to or withdrawing power from the grid when expected generation does not equal expected load from minute to minute.[97] Further, it allows grid operators and utilities to reduce peak electricity periods and avoid the need to build new generation capacity or construct new transmission lines to ensure adequate electricity resources in times of high demand.[98] As Professor Amy Stein explains, energy storage also increases the efficiency of electricity generation by reducing times of over-generation, such as during the night when wind generation is strongest but electricity demand is weakest.[99] Today, without significant energy storage, grid operators are forced to "curtail" those wind energy resources to avoid an oversupply of electricity, but utilities must often still pay for the unused wind generation under their contracts with wind operators.[100] Finally, Professor Stein notes that energy storage can have significant environmental benefits because it allows a greater reliance on renewable energy resources such as wind and solar energy. Energy storage allows for electricity generated from those resources to be temporarily held at a facility and used later, thus alleviating the intermittency challenges associated with those resources.[101]

Energy storage can be used at the utility-scale generation level, at the transmission level, and at the distribution or microgrid level.[102] Several emerging projects demonstrate these benefits:

> In late 2014, Oncor—Texas's largest transmission line network operator—announced that it would seek regulatory approval to invest billions of dollars in utility-scale batteries beginning in 2018, which would allow it to store electricity at night, when demand—and cost—is lowest and also when wind energy is at its peak. Oncor has

---

[96] Stein, *supra* note 95, at 710.

[97] *Id.* at 710–11.

[98] *See* SETH MULLENDORE, ENERGY STORAGE AND ELECTRICITY MARKETS (Resilient Power Aug. 2015) (discussing potential benefits of energy storage); *see also* Shelley Welton, *Non-Transmission Alternatives*, 39 HARV. ENVTL. L. REV. 457, 467 (2015) (discussing potential of energy storage to replace new electricity transmission by storing off-peak energy and releasing it during peak demand, serving as a balancing resource for distributed generation, and even serving as the transmission function itself).

[99] Stein, *supra* note 95, at 713–14.

[100] *Id.* at 713.

[101] *Id.* at 715.

[102] NAT'L RENEWABLE ENERGY LAB., ISSUE BRIEF: A SURVEY OF STATE POLICIES TO SUPPORT UTILITY-SCALE AND DISTRIBUTED-ENERGY STORAGE (Sept. 2014), http://www.nrel.gov/docs/fy14osti/62726.pdf.

also begun micro-grid pilot projects using battery technology, solar energy, and back-up generators. In other parts of the country, California has placed mandates on utilities to produce 1.3 gigawatts of energy storage by 2022, along with additional regional procurements. Hawaii and New York also have multiple storage pilots and projects underway.[103]

Although Oncor's utility-scale battery initiative has faced barriers under Texas law, thus stalling the project, similar projects around the country are likely.[104] Moreover, Tesla Motors has expanded on its battery technology in the EV market and has begun to offer utility-scale, commercial, and residential lithium-ion batteries for energy storage. In 2015, Tesla announced the release of the "Powerwall," designed for homeowners with PV rooftop solar panels in a 7-kWh daily battery.[105] Tesla also offers a larger "Powerpack" 100-kWh battery tower, which allows larger consumers and electricity providers, such as businesses and utilities, to store electricity.[106] Just as batteries can provide backup power when certain types of generation decline, natural gas turbines can serve a similar function.[107] For example, at one location in Florida, a solar plant is co-located with a natural gas-fired plant, and the natural gas portion of the plant can run during nighttime hours and on cloudy days.[108]

Energy storage is closely related to the increased use of renewable energy in other ways, too. For instance, as California attempts to implement its 50 percent renewable portfolio standard, many in the state are concerned about a mismatch between times of peak solar generation (from mid-morning and mid-afternoon) and peak wind generation (at night) as compared to peak electric demand

---

[103] Klass, *supra* note 92 at 17.

[104] *See* Gavin Bade, *Whatever Happened to Oncor's Big Energy Storage Plans?*, UTILITY DIVE, Sept. 1, 2015 (explaining that Oncor, as a regulated transmission and distribution utility, cannot participate in energy generation markets under Texas law, and that Oncor's proposal to enhance renewable energy storage is considered to be on the generation side of the line rather than the transmission and distribution side of the line).

[105] Ann C. Mulkern, *Tesla Says it Will Double Capacity of Home Battery*, ENERGYWIRE, June 10, 2015; *see also* Julia Pyper, *Tesla Discontinues 10 kWh Home Battery*, GREENTECH MEDIA, Mar. 18, 2016 (reporting that Tesla found a better market for the 7kW daily use battery than the 10 kW battery for longer-term storage of PV solar energy).

[106] Klass, *supra* note 92; Mulkern, *supra* note 105.

[107] Klass, *supra* note 92.

[108] Angela Neville, *Top Plant: Martin Next Generation Solar Energy Center, Indiantown, Martin County, Fl.*, POWER, Dec. 1, 2011, http://www.powermag.com/top-plantmartin-next-generation-solar-energy-center-indiantown-martin-county-florida/.

times in the late afternoon or early evening.[109] Without energy storage, there will be an oversupply of electricity during mid-day and at night, and fossil fuel generation that can ramp up quickly must be used for the late afternoon and early evening demand times. Using energy storage, utilities and regulators can "smooth out" the supply and demand curve for electricity (also known as the "duck curve" because of its shape), resulting in less wasted energy and more efficient use of renewable energy resources.[110]

V. Microgrids

"Microgrids" are "localized grids that can disconnect from the traditional grid to operate autonomously"[111] and independently from large-scale electricity, transmission, and distribution systems run by electricity generators, utilities, and grid operators.[112] Microgrids work by using distributed (i.e., on-site) generation technologies such as solar PV, on-site natural gas or diesel generators, combined heat and power (CHP) generation, or fuel cells.[113] A microgrid may be configured to be linked with the macrogrid but have the ability to "island" in cases of grid outages or, in the alternative, be completely independent from it. Microgrids, when powered by PV solar, can promote increased use of renewable energy resources, and all types of microgrids can help meet grid resiliency goals, reduce large-scale power outages, provide demand response, and promote greater use of energy storage. Microgrids can also reduce energy losses in transmission and distribution because the energy need not travel long distances from generation to consumption.[114]

There has been a growing interest in the development of microgrids as increasingly severe weather events have posed greater

---

[109] *See, e.g.,* PAUL DENHOLM, ET AL., NAT'L RENEWABLE ENERGY LAB., OVERGENERATION FROM SOLAR ENERGY IN CALIFORNIA: A FIELD GUIDE TO THE DUCK CHART (Nov. 2015).

[110] *Id.*

[111] *The Role of Microgrids in Helping to Advance the Nation's Energy System*, U.S. DEP'T OF ENERGY, OFFICE OF ELECTRICITY DELIVERY & ENERGY RELIABILITY, http://energy.gov/oe/services/technology-development/smart-grid/role-microgrids-helping-advance-nation-s-energy-system.

[112] *How Microgrids Work*, U.S. DEP'T OF ENERGY, http://www.energy.gov/articles/how-microgrids-work; *Examples of Microgrids*, U.S. DEP'T OF ENERGY, BERKELEY LAB, https://building-microgrid.lbl.gov/examples-microgrids (giving numerous examples of distributed energy resource networks that act as microgrids in cities, such as Fort Collins, Colorado, universities, such as New York University, and other government and private institutions using solar, natural gas, and steam powered on-site generation resources).

[113] *See The Role of Microgrids in Helping to Advance the Nation's Energy System*, U.S. DEP'T OF ENERGY, ENERGY.GOV, http://energy.gov/oe/services/technology-development/smart-grid/role-microgrids-helping-advance-nation-s-energy-system; Sara C. Bronin, *Curbing Energy Sprawl with Microgrids*, 43 CONN. L. REV. 547, 559 (2010).

[114] *See* Bronin, *supra* note 113, at 548.

threats to the macrogrid. For instance, when Superstorm Sandy struck the U.S. east coast in October 2012, 8.5 million residential, commercial, and industrial electricity customers in 17 states lost power.[115] Although most of these customers had power restored within a couple of weeks, approximately 1 percent still lacked electricity at the end of November 2012. The disruptions to the grid resulted in billions of dollars in economic losses, not to mention significant harm to human health.[116] Notable exceptions to the widespread power losses were Princeton University, parts of New York University in lower Manhattan, several hospitals, and a few residential housing complexes. All of these facilities had on-site combined heat and power generation facilities that were able to disconnect from the grid and provide continuous power during the disaster.[117]

As a result of the potential benefits of microgrids, particularly during weather-related and other disasters, the federal government and states are increasingly supporting microgrid projects through incentives, research, and funding. As the Congressional Research Service explains, the U.S. Department of Defense has installed "a few pilot microgrid projects with renewable electricity and energy storage to test the economics, resilience, and operational independence in the event of a large-scale power outage."[118] Many states have pilot projects and incentive programs designed to encourage research and investment in microgrids.[119] As described in more detail in Chapter 5, numerous states have also created net metering and other programs to incentivize electricity customers and utilities to increase the use of PV solar panels, which would allow renewable energy to play a more central role in microgrid development.

Although there are significant potential benefits associated with microgrids, it is important to keep in mind that most existing microgrids in the U.S. are powered by natural gas or diesel fuel rather than renewable energy.[120] As a result, the resiliency and independence associated with microgrids may be offset by the continued reliance on fossil fuels, particularly if microgrids are used

---

[115] *Electricity Restored to Many in Northeast But Outages Persist*, U.S. ENERGY INFO. ADMIN., Nov. 9, 2012, http://www.eia.gov/todayinenergy/detail.cfm?id=8730.

[116] *Id.*; DAVIES ET AL., *supra* note 8, at 19 (citing statistics).

[117] *See, e.g.*, James Van Nostrand, *Keeping the Lights on During Superstorm Sandy, Climate Change Adaptation and the Resiliency Benefits of Distributed Generation*, 23 N.Y.U. ENVTL. L. REV. 93, 107–11 (2015).

[118] *See* RICHARD J. CAMPBELL, CONG. RESEARCH SERV., CUSTOMER CHOICE AND THE POWER INDUSTRY OF THE FUTURE 12 (Sept. 22, 2014).

[119] *See* KEMA, INC., MICROGRIDS—BENEFITS, MODELS, BARRIERS, AND SUGGESTED POLICY INITIATIVES FOR THE COMMONWEALTH OF MASSACHUSETTS (Feb. 3, 2014).

[120] MATT GRIMLEY & JOHN FARRELL, INST. FOR LOCAL SELF RELIANCE, MIGHTY MICROGRIDS 7 (2016).

as a substitute for increases in utility-scale wind or solar energy in the macrogrid.[121] Nevertheless, there are good arguments that increased development of microgrids with a greater focus on use of renewable-powered distributed generation and demand response can meet the resiliency, economic, and efficiency goals discussed above while also supporting greater use of utility-scale renewable energy in the macrogrid.[122]

Nevertheless, increasing use of microgrids and renewable energy calls into question the regulatory model that has supported vertically integrated utilities for over a century. Even as FERC and state legislatures have introduced competition in wholesale electricity generation, and some states have created competition in retail electricity markets, there has been little change in the structure that compensates utilities based on the amount of electricity they provide through the macrogrid. Indeed, one consulting firm anticipates that conventional power plant operators may lose approximately $2 billion in annual revenue starting in 2019 as increases in residential PV solar installations decrease the demand on the traditional electric grid.[123] It reports that more than one million U.S. homes will have solar panels by May 2016, which will cause a projected decline of 1,400 megawatts of electricity demanded from power plants in 2019.[124] This has a major financial impact on independent generators as well as public utilities with substantial generation capacity in their portfolios.

As a result, policymakers, experts, and industry representatives increasingly express concern regarding a "utility death spiral," as more customers generate their own electricity at the same time that electricity demand overall has remained flat, preventing utilities from achieving economies of scale through growth as was done for decades. This has led experts to encourage the use of new approaches to utility ratemaking and compensation that are tied to electricity *services* rather than electricity *generation*. These approaches includes compensating utilities for increasing energy efficiency and

---

[121] *See, e.g.*, MATTHEW CHRISTIANSEN & ELIZABETH B. STEIN, NYU LAW, THE RISE OF DG: OPTIONS FOR ADDRESSING THE ENVIRONMENTAL CONSEQUENCES OF INCREASED DISTRIBUTED GENERATION (Feb. 2016) (discussing incentives for distributed generation and potential adverse environmental consequences when such generation is powered by fossil fuels).

[122] CHRISTOPHER VILLARREAL, DAVID ERICKSON, & MARZIA ZAFAR, CALIFORNIA PUBLIC UTILITIES COMM'N, MICROGRIDS: A REGULATORY PERSPECTIVE 9–11 (Apr. 14, 2014) (discussing how expansion of microgrids can support macrogrid operations and efficiencies, provide additional resilience in times of macrogrid outages from storms or other electric disruptions, and can effectively manage intermittency of microgrid and macrogrid renewable energy resources by coordinating storage and demand with generator output).

[123] *Utilities to Lose $2B to Rooftop Panels in 2019*, GREENWIRE, Mar. 21, 2016.

[124] *Id.*

conservation at the wholesale and retail levels, implementing smart meter and smart grid programs for customers, and facilitating increased use of solar PV and other distributed generation powered by renewable energy.[125]

---

[125] *See, e.g.*, Edison Elec. Inst. & Nat. Res. Defense Fund, Joint Statement to State Utility Regulators, Feb. 12, 2014; E21 INITIATIVE, PHASE I REPORT: CHARTING A PATH TO A 21ST CENTURY ENERGY SYSTEM IN MINNESOTA (Dec. 2014); William Boyd, *Public Utility and the Low-Carbon Future*, 61 UCLA L. REV. 1614 (2014) (providing a historical perspective to argue that the public utility can play an important role in decarbonizing the U.S. electricity system and respond to recent changes in electricity generation and distribution services); Gavin Bade, *New York PSC Enacts New Revenue Model for Utilities in REV Proceeding*, UTILITY DIVE, May 20, 2016.

TABLE OF CASES

Aiken Cty., In re, 138
Alliance to Save the Mattaponi v. U.S. Army Corps, 35
American Fuels & Petrochemical Ass'n v. Corey, 119
American Petroleum Institute v. EPA, 157
Bragg v. W. Va. Coal Ass'n, 101
Butler v. Charles Powers Estate, 48
California ex rel. Lockyer v. Fed. Energy Regulatory Comm'n, 167, 180
California Wilderness Coal. v. U.S. Dept. of Energy, 21
Chevron U.S.A., Inc. v. NRDC, 12
Citizens to Preserve Overton Park v. Volpe, 11
Coastal Oil & Gas Corp. v. Garza Energy Trust, 50
Colorado Interstate Gas Co. v. Wright, 93
Dabney-Johnston Oil Corp. v. Walden, 49
EarthReports, Inc. v. Fed. Energy Regulatory Comm'n, 32
Ely v. Cabot Oil and Gas Corp., 70
Energy Mgmt. Corp. v. City of Shreveport, 68
Entergy Louisiana, Inc. v. Louisiana Public Service Comm'n, 192
Entergy Nuclear Vermont Yankee v. Shumlin, 136, 188
EPA v. EME Homer Generation, 122
Fed. Energy Regulatory Comm'n v. Elec. Power Supply Ass'n, 178, 193, 215
Fed. Power Comm'n v. Florida Power & Light, 188
Fed. Power Comm'n v. Sierra Pac. Power Co., 174
Fort Collins, City of v. Colo. Oil and Gas Ass'n., 68
FPL Farming Ltd. v. Envtl. Processing Sys., 110
Friedman v. Texaco, 48
Getty Oil Co. v. Jones, 55
Grimes v. Goodman Drilling Co., 55
Grocery Mfrs. Ass'n v. EPA, 156, 157
Gulf Restoration Network v. United Bulk Terminals, 39
Hughes v. Talen Energy, 194
Illinois Commerce Comm'n v. Fed. Energy Regulatory Comm'n, 18, 19, 80, 150
Jefferson County PUD v. Washington Dep't. of Ecology, 37
Ladra v. New Dominion, LLC, 70, 110
Larson v. Sinclair Transportation Co., 23, 24
Longmont, City of v. Colo. Oil and Gas Ass'n, 68
Lujan v. National Wildlife Federation, 10
Massachusetts v. EPA, 10, 124
Michigan v. EPA, 121, 123
Mingo Logan Coal Co. v. EPA, 104
Minisink Residents for Envtl. Pres. & Safety v. Fed. Energy Regulatory Comm'n, 27
Missouri, State of ex rel. Barrett v. Kansas Natural Gas. Co., 181
Monroe Energy, LLC v. EPA, 157
Morgan Stanley Capital Group v. Public Utility District No, 174
Morrison, State ex rel. v. Beck Energy Corp., 68
Motor Vehicle Manufacturers Ass'n v. State Farm Mut. Auto Ins. Co., 11
Nantahala Power & Light Co. v. Thornburg, 192
National Petrochemical & Refiners Ass'n v. EPA, 158
New Jersey v. EPA, 123
New York v. Fed. Energy Regulatory Comm'n, 18, 189
New York v. U.S. Army Corps of Eng'rs, 108
North Carolina v. EPA, 122
North Dakota v. Heydinger, 129
Norton v. Southern Utah Wilderness Alliance, 11
Otter Tail Power Co. v. United States, 175
Otter Tail Power Co., United States v., 175

Pacific Gas & Elec. Co. v. State Energy Res. Conservation & Dev. Comm'n, 136
Parr v. Aruba Petroleum, 70
Pauly v. Bethenergy Mines, Inc., 12
Permian Basin Area Rate Cases, In re, 183
Phillips Petroleum Co. v. Wisconsin, 166
Piedmont Envtl. Council v. Fed. Energy Regulatory Comm'n, 21
Public Utilities Commission v. Attleboro Steam and Electric Co., 170
Public Utility District No. 1 of Snohomish Cty. v. Dynegy Power Marketing, Inc., 180
Robinson Township v. Commonwealth of Pennsylvania, 68
Rocky Mountain Farmers Union v. Corey, 116
Sabine Pass Liquefaction, LLC, In re, 31
Sierra Club v. Fed. Energy Regulatory Comm'n, 32
Sierra Club v. Morton, 10
Stephens County v. Mid-Kansas Oil & Gas Co., 49
Stone v. Chesapeake Appalachia, LLC, 50
Sun Oil Co. v. Whitaker, 55
SWEPI v. Mora County, 68
Tampa Electric Co. v. Garcia, 193
Ten Taxpayer Citizens Grp. v. Cape Wind Assoc., 14
Texas Commercial Energy v. TXU Energy, 173
Texas Oil and Gas Ass'n v. City of Denton, 67
Thompson v. Heineman, 26
United Gas Pipe Line Co. v. Mobile Gas Serv. Corp., 174
Wallach v. Town of Dryden, 68
West Virginia v. EPA, 126
White Stallion Energy Ctr. v. EPA, 124
Wyoming v. U.S. Dept. of the Interior, 64, 106
Zimmerman v. Bd. of County Comm'rs, 16

INDEX

References are to Pages

ABANDONED MINE RECLAMATION
Environmental regulations, 101

ACCOMMODATION DOCTRINE
Oil and gas extraction, disputes between service and mineral owners, 55

ACID RAIN
Environmental regulations, 120

ADMINISTRATIVE LAW
Generally, 7–12
Adjudications, 9–10
Administrative Procedure Act, 8
Formal v. informal rulemaking, 9
Judicial review, 10–12
Rulemaking, 8–9

AIR POLLUTION
Environmental Regulations, this index

ALISO CANYON NATURAL GAS STORAGE FACILITY
Gas leak, 92

AMERICAN RECOVERY AND REINVESTMENT ACT
Smart grid and distributed generation, 215

APPEAL AND REVIEW
Administrative law, 10–12
Federal Energy Regulatory Commission, 11
Natural Gas Act, 10–11

ARAB OIL EMBARGO
Generally, 1

ARMY CORPS OF ENGINEERS
Coal transportation, 35–38

ATOMIC ENERGY
Nuclear Energy, this index

ATOMIC ENERGY ACT
Generally, 4

ATOMIC ENERGY COMMISSION
Environmental regulations, 134, 136

BALD AND GOLDEN EAGLE PROTECTION ACT
Oil and gas extraction, 44

BATTERY TECHNOLOGY
Electric vehicles, 222

BIOFUELS
Environmental Regulations, this index
Mandates, this index
Tax incentives, 147

BLACKOUTS
Transportation of energy, 86

BONNEVILLE POWER ADMINISTRATION
Transportation of energy, 83

BROWNOUTS
Transportation of energy, 86

BUREAU OF LAND MANAGEMENT
Coal extraction, environmental regulations, 99
Hydraulic fracturing, 63–64
Oil and gas extraction, 43, 56–57

BUREAU OF OCEAN ENERGY MANAGEMENT
Oil and gas extraction, 59–61

CALIFORNIA GLOBAL WARMING SOLUTIONS ACT
Electricity, 128–129

CAP-AND-TRADE PROGRAM
Electricity, environmental regulations, 120–121

CAPTURE RULE
Oil and gas extraction, onshore mineral ownership and leasing, 50–53

CARBON MONOXIDE
Environmental regulations, state emission standards, 127

CASH GRANT PROGRAM
Tax incentives, renewable energy projects, 145–146

CERTIFICATE OF NEED OR CERTIFICATE OF PUBLIC CONVENIENCE AND NECESSITY
Siting of electric transmission lines, 18

CHARGING STATIONS
Electric vehicles, 222

CLEAN AIR ACT
Generally, 1, 4
Coal transportation, 37
Electricity, 120–123
Oil and gas extraction and
    operations, 43–44, 105–107

CLEAN AIR INTERSTATE RULE
Electricity, 122

CLEAN AIR MERCURY RULE
Electricity, 123

CLEAN LINE ENERGY PARTNERS
Electric transmission lines, siting, 20

CLEAN POWER PLAN
Electricity, 125–127

CLEAN RENEWABLE ENERGY BONDS
Tax incentives, 146–147

CLEAN WATER ACT
Generally, 1, 5
Coal transportation, 35
Oil and gas extraction, 43–45

COAL
Clean Air Act, 37
Clean Water Act, 35
Coastal Zone Management Act, 37
Environmental Regulations, this index
National Environmental Policy Act, 35–36
Siting of Energy Facilities, this index
Tax incentives, coal bed methane gas, 142

COASTAL ZONE MANAGEMENT ACT
Generally, 5
Coal transportation, 37
Oil and gas extraction, 60

COMMON LAW
Generally, 1
Hydraulic fracturing, 70
Oil and gas operations,
    environmental regulations, 109–110

COMMUNITY KNOWLEDGE TEST
Oil and gas extraction, onshore mineral ownership and leasing, 47–48

COMPREHENSIVE ENVIRONMENTAL RESPONSE, COMPENSATION AND LIABILITY ACT
Oil and gas extraction and
    operations, 43, 45, 105–107

CONDEMNATION
Siting of oil pipelines, 23–24

CONTIGUOUS WATERS
Oil and gas extraction, offshore leasing and regulation, 58

CORPORATE AVERAGE FUEL ECONOMY STANDARDS
Transportation, environmental regulations, 111–113

CROSS-STATE AIR POLLUTION RULE
Electricity, 122

CYBERSECURITY
Generally, 4

DATA PRIVACY
Smart grid and distributed generation, 216–217

DATA STANDARDIZATION
Smart grid and distributed generation, 216–217

DELAWARE RIVER BASIN COMMISSION
Generally, 7
Hydraulic fracturing, 69
Oil and gas operations,
    environmental regulations, 107–108

DEMAND RESPONSE RESOURCES
Markets, conflict between state and federal jurisdiction, 193–194

DEPARTMENT OF ENERGY
Generally, 6
LNG terminals, 29
Loan guarantee program, renewable energy projects, 146
Nuclear power, environmental regulations, 134–135

DEPARTMENT OF THE INTERIOR
Generally, 6

DEPARTMENT OF TRANSPORTATION
Generally, 6

INDEX

DISTRIBUTED GENERATION
Smart Grid and Distributed
    Generation, this index

EARTHQUAKES
Hydraulic fracturing, 61–62, 67

ELECTRICITY
Cybersecurity, this index
Environmental Regulations, this
    index
Mandates, this index
Markets, this index
Siting of Energy Facilities, this index

ELECTRIC GRID
Computer hacking, 87
Critical cyber assets, 90
Physical Security Reliability
    Standard, 89

ELECTRIC VEHICLES
Smart Grid and Distributed
    Generation, this index

EMINENT DOMAIN
Siting of oil pipelines, 23–24

ENDANGERED SPECIES ACT
Generally, 5
Oil and gas extraction and
    operations, 43, 105

ENERGY EFFICIENCY POLICIES
Electricity, 127–128
Mandates, energy efficiency resource
    standards, 155, 219
Smart grid and distributed
    generation, resource standards,
    219

ENERGY INDEPENDENCE AND SECURITY ACT
Generally, 4
Biofuels, 114
Mandates, 155
Smart grid and distributed
    generation, 214

ENERGY POLICY ACT
Generally, 4
Electricity markets, federal
    regulation of, 176
Energy transmission lines, siting, 20
Smart grid and distributed
    generation, 213

ENERGY POLICY CONSERVATION ACT
Imports and exports, 200

ENVIRONMENTAL PROTECTION AGENCY
Generally, 6

Hydraulic fracturing, 65
Transportation, 111–113

ENVIRONMENTAL REGULATIONS
Generally, 97–138
Abandoned mine reclamation, 101
Acid rain, 120
Air pollution
    Electricity, below
    Transportation, below
Atomic Energy Commission, 134, 136
Biofuels
    Generally, 113–119
    Energy Independence and
        Security Act, 114
    Ethanol, 114–118
    Low Carbon Fuel Standard,
        116–119
    Renewable fuel standards, 114–
        117
Bureau of Land Management, coal
    extraction, 99
California Global Warming Solutions
    Act, electricity, 128–129
Cap-and-trade program, electricity,
    120–121
Carbon monoxide, state emission
    standards, 127
Clean Air Act, this index
Clean Air Interstate Rule, electricity,
    122
Clean Air Mercury Rule, electricity,
    123
Clean Power Plan, electricity, 125–
    127
Coal extraction
    Generally, 97–104
    Abandoned mine reclamation,
        101
    Bureau of Land Management,
        99
    Federal Mine Safety and
        Health Review
        Commission, 98
    Mine Improvement and New
        Energy Response Act, 98
    Mineral Leasing Act, 99
    Mine Safety and Health Act, 98
    Mountaintop mining, 102–104
    Office of Surface Mining
        Reclamation and
        Enforcement, 100–101
    Surface Mining Control and
        Reclamation Act, 98–104
Coal-fired generation of electricity,
    state bans on new, 129
Comprehensive Environmental
    Response, Compensation and
    Liability Act, oil and gas
    operations, 105–107

INDEX

Corporate Average Fuel Economy standards, transportation, 111–113
Cross-State Air Pollution Rule, electricity, 122
Delaware River Basin Commission, oil and gas operations, 107–108
Department of Energy, nuclear power, 134–135
Disposal of nuclear waste, 137–138
Electricity
    Generally, 119–130
    Acid rain, 120
    Air pollution, generally, 119–130
    California Global Warming Solutions Act, 128–129
    Cap-and-trade program, 120–121
    Carbon monoxide, state emission standards, 127
    Clean Air Act, 120–123
    Clean Air Interstate Rule, 122
    Clean Air Mercury Rule, 123
    Clean Power Plan, 125–127
    Coal-fired generation, state bans on new, 129
    Cross-State Air Pollution Rule, 122
    Energy efficiency policies, state standards, 127–128
    Federal statutes and regulations, 119–127
    Greenhouse gas emissions, 124 et seq.
    Hazardous air pollutants, 122–123
    Maximum achievable control technology, 123
    Mercury pollution, 122
    Nitrogen oxide emissions, 120–122
    Regional collaborations, 129–130
    Renewable portfolio standards, state standards, 128
    State implementation plans, 121–122
    State statutes and regulations, 127–129
    Sulfur dioxide emissions, 120–121
Endangered Species Act, oil and gas operations, 105
Energy efficiency policies, electricity, 127–128
Energy Independence and Security Act, biofuels, 114
Environmental Protection Agency, transportation, 111–113
Ethanol, biofuels, 114–118

Federal Land Policy and Management Act, oil and gas operations, 106
Federal Mine Safety and Health Review Commission, 98
Gas operations. Oil and gas operations, below
Greenhouse gas emissions
    Electricity, 124 et seq.
    Regional collaborations, 129–130
    Transportation, 110 et seq.
Hazardous air pollutants, electricity, 122–123
Hydraulic fracturing, oil and gas operations, 105, 107
Hydropower
    Generally, 130–134
    Conventional hydropower, 130–134
    Licenses, 132–133
    Public v. private sector facilities, 130–131
Licenses, hydropower, 132–133
Low Carbon Fuel Standard, 116–119
Maximum achievable control technology, electricity, 123
Mercury pollution, electricity, 122
Mine Improvement and New Energy Response Act, 98
Mineral Leasing Act, 99
Mine Safety and Health Act, 98
Mountaintop mining, 102–104
National Environmental Policy Act, this index
New motor vehicle regulations, 111–113
Nitrogen oxide emissions, electricity, 120–122
Nuclear power
    Generally, 134–138
    Atomic Energy Commission, 134, 136
    Department of Energy, 134–135
    Disposal of nuclear waste, 137–138
    Nuclear Regulatory Commission, 134–135
    Permits, 135–136
Occupational Safety and Health Act, oil and gas operations, 105
Office of Surface Mining Reclamation and Enforcement, 100–101
Oil and gas operations
    Generally, 105–110
    Clean Air Act, 105–107
    Comprehensive Environmental Response, Compensation and Liability Act, 105–107

Delaware River Basin
    Commission, 107–108
Endangered Species Act, 105
Federal Land Policy and
    Management Act, 106
Hydraulic fracturing, 105, 107
Occupational Safety and
    Health Act, 105
Oil Pollution Act, 105, 107
Resource Conservation and
    Recovery Act, 106
Safe Drinking Water Act, 105,
    107
Susquehanna River Basin
    Commission, 108–109
Tort claims and other common
    law claims, 109–110
Water pollution, 105
Oil Pollution Act, 105, 107
Permits, nuclear power, 135–136
Regional collaborations
Electricity, 129–130
Greenhouse gas emissions, 129–130
Renewable fuel standards, 114–117
Renewable portfolio standards,
    electricity, 128
Resource Conservation and Recovery
    Act, oil and gas operations, 106
Safe Drinking Water Act, oil and gas
    operations, 105, 107
State implementation plans,
    electricity, 121–122
State statutes and regulations,
    electricity, 127–129
Sulfur dioxide emissions, electricity,
    120–121
Surface Mining Control and
    Reclamation Act, 98–104
Susquehanna River Basin
    Commission, oil and gas
    operations, 108–109
Tort claims and other common law
    claims, oil and gas operations,
    109–110
Transportation
    Generally, 110–119
    Air pollution, generally, 110–119
    Biofuels, above
    Corporate Average Fuel
        Economy standards, 111–113
    Environmental Protection
        Agency, 111–113
    Greenhouse gas emissions, 110
        et seq.
    New motor vehicle regulations,
        111–113
    Vehicle emissions, 111–113
Vehicle emissions, 111–113

Water pollution, oil and gas
    operations, 105

ETHANOL
Biofuels, 114–118
Tax credits, 140–141

EXCLUSIVE ECONOMIC ZONES
Oil and gas extraction, offshore
    leasing and regulation, 58

**EXEMPT WHOLESALE
GENERATORS**
Electricity markets, federal
    regulation of, 179

EXPORTS
Imports and Exports, this index

**FEDERAL ENERGY
REGULATORY
COMMISSION**
Generally, 5–7
Energy transmission lines, siting,
    20–22
Judicial review, 11
LNG terminals, 29–32
Markets, this index
Natural gas pipelines, 27–28
Transportation of Energy, this index

**FEDERAL LAND POLICY AND
MANAGEMENT ACT**
Oil and gas extraction and
    operations, 43, 56, 106

**FEDERAL MINE SAFETY AND
HEALTH REVIEW
COMMISSION**
Environmental regulations, 98

**FEDERAL OIL AND GAS
ROYALTY MANAGEMENT
ACT**
Oil and gas extraction, 43

FEDERAL POWER ACT
Generally, 1, 4
Electricity markets, 171
Electric transmission lines, siting, 18
Judicial review, 10–11
Transportation of energy, 73–74

**FEDERAL POWER
COMMISSION**
Generally, 5
Electricity markets, 171

**FEDERAL RAILROAD
ADMINISTRATION**
Transportation of energy, safety, 94–95

FEED-IN TARIFFS
Mandates, 152–153

FINANCING
Generally, 139–140

FINANCING OF RENEWABLE ENERGY PROJECTS
Generally, 158–164
Distributed infrastructure
    Generally, 161–164
    Leases, 164
    Property Assessed Clean Energy, 162–163
    Third-party ownership, 163–164
Leases, distributed infrastructure, 164
Master limited partnerships, utility-scale infrastructure, 160–161
Partnership-flips, utility-scale infrastructure, 159–160
Property Assessed Clean Energy, distributed infrastructure, 162–163
Real estate investment trusts, utility-scale infrastructure, 161
Third-party ownership, distributed infrastructure, 163–164
Utility-scale infrastructure
    Generally, 158–161
    Master limited partnerships, 160–161
    Partnership-flips, 159–160
    Real estate investment trusts, 161

FISH AND WILDLIFE SERVICE
Hydraulic fracturing, 64–65

FOSSIL FUELS
Coal, this index
Nuclear Energy, this index
Oil and Gas, this index
Tax Incentives, this index

FREE TRADE AGREEMENTS
Imports and exports, 203
LNG terminals, 29

GAS
Oil and Gas, this index

GENERATION OF POWER
Siting of Energy Facilities, this index

GREENHOUSE GAS EMISSIONS
Environmental Regulations, this index

GULF OF MEXICO ENERGY SECURITY ACT
Oil and gas extraction, 59

HAZARDOUS AIR POLLUTANTS
Electricity, environmental regulations, 122–123

HIGH-HAZARD FLAMMABLE TRAINS
Transportation of energy, 95–96

HIGH THREAD URBAN AREAS
Railroads, transportation of energy, 96

HYDRAULIC FRACTURING
Generally, 2, 41–42, 61–72
Bureau of Land Management, 63–64
Common law, 70
Delaware River Basin Commission, 69
Earthquakes, 61–62, 67
Environmental Protection Agency, 65
Environmental regulations, 105, 107
Fish and Wildlife Service, 64–65
Lending agreements, 71
LNG terminals, 30–32
Local government regulations, 67–69
Memoranda of understanding with local governments, 71
Nuisance, 70
Private law, 71–72
Regional government agency regulations, 69
Regulation of development, 63–69
Safe Drinking Water Act, exemption, 65, 107
State moratoriums, 66
State v. federal regulations, 64–67
Susquehanna River Basin Commission, 69
Tax incentives, 142
Trespass, 70

HYDROPOWER
Environmental Regulations, this index

IMPORTS AND EXPORTS
Generally, 197–208
Crude oil exports, 199–202
Energy Policy Conservation Act, 200
Federal regulation, 199–208
Free trade agreements, 203
Liquefied natural gas exports, 200, 202–208
National interest determination, 200–202
Organization of Arab Petroleum Exporting Countries, 197
Siting of coal transportation facilities, 32–34
Trends, 198–199

INDEPENDENT SYSTEM OPERATORS
Generally, 7
Electricity markets, federal regulation of, 168, 177–178, 180

Transportation of energy, 75–77

INTRASTATE TRANSMISSION LINES
Electricity markets, state regulation of, 188–190

INVESTMENT TAX CREDIT
Renewable energy projects, 145

KEYSTONE XL PIPELINE
Siting of energy facilities, 24–27

LARGE-SCALE BATTERIES
Smart grid and distributed generation, 225–226

LEASES
Financing of renewable energy projects, distributed infrastructure, 164

LICENSES
Hydropower, 132–133

LNG EXPORTS
Imports and exports, 200, 202–208

LNG TERMINALS
Department of Energy, 29
Federal Energy Regulatory Commission, 29–32
Hydraulic fracturing, 30–32
Natural Gas Act, 29–30
Siting of energy facilities, 28–32

LOAD CENTERS
Siting of electric transmission lines, 18–19

LOCAL DISTRIBUTION COMPANIES
Natural gas markets, federal regulation of, 168

LOW CARBON FUEL STANDARD
Environmental regulations, 116–119

MANDATES
Generally, 139–140, 149–157
Biofuels, generally, 155–157
Electricity, generally, 149–155
Energy efficiency resource standards, 155, 219
Energy Independence and Security Act, 155
Feed-in tariffs, 152–153
Net metering, 153–155
Renewable energy credits, 150–152
Renewable fuel standards, 155–157
Renewable portfolio standards, 149–150

MARKETS
Generally, 165–195
Conflict between state and federal jurisdiction, 192–195
Demand response resources, conflict between state and federal jurisdiction, 193–194
Electricity markets, federal regulation
Generally, 167–181
Energy Policy Act, 176
Exempt wholesale generators, 179
Federal Energy Regulatory Commission, 171 et seq.
Federal Power Act and Federal Power Commission, 171
Independent system operator's, 168, 177–178, 180
Open access transmission tariffs, 176
Power purchase agreements, 168
Public Utility Holding Company Act, 171, 179
Public Utility Regulatory Policies Act, 178–179
Qualifying small power production facilities, 178–179
Rates, 171–174
Regional transmission organizations, 168, 177–178, 180
"Wheeling," 174–176
Electricity markets, state regulation
Generally, 187–191
Distribution to retail customers, 190–191
Intrastate transmission lines, 188–190
Retail sales, 187–188
Energy Policy Act, federal regulation of electricity markets, 176
Exempt wholesale generators, federal regulation of electricity markets, 179
Federal Energy Regulatory Commission
Electricity markets, 171 et seq.
Natural gas markets, 182
Oil markets, 184–186
Federal Power Act and Federal Power Commission, electricity markets, 171
Federal regulation, generally, 167–186, 192–195
Generation capacity, conflict between state and federal jurisdiction, 194–195

242 INDEX

Independent system operators, federal regulation of electricity markets, 168, 177–178, 180
Intrastate transmission lines, state regulation of electricity markets, 188–190
Local distribution companies, federal regulation of natural gas markets, 168
Natural Gas Act, federal regulation of natural gas markets, 182
Natural gas markets, federal regulation
    Generally, 168–169, 181–184
    Federal Energy Regulatory Commission, 182
    Local distribution companies, 168
    Natural Gas Act, 182
    Pipelines, 182–183
    Rates, 183
Natural gas markets, state regulation, 191–192
Natural monopolies, status as, 165–167
Oil markets, federal regulation
    Generally, 169, 184–186
    Federal Energy Regulatory Commission, 184–186
    Pipelines, 184–186
    Railroad transportation, 186
Oil markets, state regulation, 191–192
Open access transmission tariffs, federal regulation of electricity markets, 176
Pipelines
    Natural gas markets, federal regulation, 182–183
    Oil markets, federal regulation, 184–186
Power purchase agreements, federal regulation of electricity markets, 168
Public Utility Holding Company Act, federal regulation of electricity markets, 171, 179
Public Utility Regulatory Policies Act, federal regulation of electricity markets, 178–179
Qualifying small power production facilities, federal regulation of electricity markets, 178–179
Railroad transportation, federal regulation of oil markets, 186
Rates
    Conflict between state and federal jurisdiction, 192–193

    Electricity markets, federal regulation, 171–174
    Natural gas markets, federal regulation, 183
Regional transmission operators, federal regulation of electricity markets, 168, 177–178, 180
Retail sales, state regulation of electricity markets, 187–188
State regulation, generally, 187–195
"Wheeling," federal regulation of electricity markets, 174–176

MASTER LIMITED PARTNERSHIPS
Financing of renewable energy projects, utility-scale infrastructure, 160–161

MAXIMUM ACHIEVABLE CONTROL TECHNOLOGY
Electricity, environmental regulations, 123

MERCURY POLLUTION
Electricity, environmental regulations, 122

MICROGRIDS
Smart grid and distributed generation, 210, 228–231

MIDCONTINENT INDEPENDENT SYSTEM OPERATOR
Electric transmission lines, siting, 19
Transportation of energy, 81, 83–84

MIGRATORY BIRD TREATY ACT
Oil and gas extraction, 44

MINE IMPROVEMENT AND NEW ENERGY RESPONSE ACT
Generally, 98

MINERAL LEASING ACT
Environmental regulations, 99
Oil and gas extraction, 42–43

MINERALS MANAGEMENT SERVICE
Oil and gas extraction, offshore leasing and regulation, 59

MINE SAFETY AND HEALTH ACT
Environmental regulations, 98

MOTOR VEHICLES
Smart Grid and Distributed Generation, this index

MOUNTAINTOP MINING
Environmental regulations, 102–104

NATIONAL ENVIRONMENTAL POLICY ACT
Generally, 1, 5
Coal transportation, 35–36
Oil and gas extraction, leasing of publicly-owned minerals, 56–57

NATIONAL INTEREST
Imports and exports, 200–202

NATIONAL INTEREST ELECTRIC TRANSMISSION CORRIDORS
Energy transmission lines, siting, 21

NATURAL GAS
Federal Energy Regulatory Commission, pipelines, 27–28
Markets, this index

NATURAL GAS ACT
Generally, 4
Judicial review, 10–11
LNG terminals, 29–30
Markets, federal regulation of, 182

NATURAL MONOPOLIES
Markets, 165–167

NET METERING
Mandates, 153–155

NITROGEN OXIDE EMISSIONS
Electricity, environmental regulations, 120–122

NONRENEWABLE ENERGY
Generally, 3

NORTH AMERICAN ELECTRIC RELIABILITY CORPORATION
Generally, 7
Transportation of energy, safety, 87–88

NUCLEAR ENERGY
Generally, 1
Environmental Regulations, this index
Siting of generation facilities, 15

NUCLEAR REGULATORY COMMISSION
Generally, 6

NUISANCE
Hydraulic fracturing, 70

OCCUPATIONAL SAFETY AND HEALTH ACT
Oil and gas operations, environmental regulations, 105

OFFICE OF FOSSIL ENERGY
LNG terminals, 29

OFFICE OF SURFACE MINING RECLAMATION AND ENFORCEMENT
Environmental regulations, 100–101

OFFSHORE LEASING AND REGULATION
Oil and Gas, this index

OIL AND GAS
Generally, 41–72
Accommodation doctrine, disputes between service and mineral owners, 55
Allowable, onshore mineral ownership and leasing, 52
Bald and Golden Eagle Protection Act, 44
Bureau of Land Management, 43, 56–57
Bureau of Ocean Energy Management, 59–61
Capture rule, onshore mineral ownership and leasing, 50–53
Clean Air Act, 43–44
Clean Water Act, 43–45
Coastal Zone Management Act, 60
Community knowledge test, onshore mineral ownership and leasing, 47–48
Comprehensive Environmental Response, Compensation and Liability Act, 43, 45
Contiguous waters, offshore leasing and regulation, 58
Endangered Species Act, 43
Environmental Regulations, this index
Exclusive economic zones, offshore leasing and regulation, 58
Exports of crude oil, 199–202
Federal Lands Policy Management Act, 43, 56
Federal Oil and Gas Royalty Management Act, 43
Gulf of Mexico Energy Security Act, 59
Hydraulic Fracturing, this index
LNG, this index
Markets, this index
Migratory Bird Treaty Act, 44
Mineral Leasing Act, 42–43
Minerals Management Service, offshore leasing and regulation, 59
National Environmental Policy Act, leasing of publicly-owned minerals, 56–57
Natural Gas, this index
Offshore leasing and regulation
Generally, 57–61

Bureau of Ocean Energy
    Management, 59–61
Coastal Zone Management Act,
    60
Contiguous waters, 58
Exclusive economic zones, 58
Gulf of Mexico Energy Security
    Act, 59
Minerals Management Service,
    59
Outer Continental Shelf Lands
    Act, 57–59
Submerged Lands Act, 57–58
United Nations Convention on
    the Law of the Sea, 58
Oil Pollution Act, 43
Onshore mineral ownership and
  leasing
        Generally, 46–57
        Accommodation doctrine,
            disputes between service
            and mineral owners, 55
        Allowable or prorationing, 52
        Capture, rule of, 50–53
        Community knowledge test,
            47–48
        Conservation regulation, 51–53
        Disputes between service and
            mineral owners, 54–55
        Pooling clauses, 51–52
        Privately-owned minerals, 53–
            54
        Publicly-owned minerals, 55–57
        Royalties, 48
        Surface destruction test, 47–48
        Transfer or reservation of
            minerals, 47
        Types of mineral ownership,
            47–50
        Unitization, 52
Outer Continental Shelf Lands Act,
    42, 57–59
Pooling clauses, onshore mineral
    ownership and leasing, 51–52
Prorationing, onshore mineral
    ownership and leasing, 52
Reserved authority of states, 45–46
Royalties, onshore mineral ownership
    and leasing, 48
Safe Drinking Water Act, 43, 45–46
Siting of Energy Facilities, this index
State v. federal authority
        Generally, 42–46
        Bald and Golden Eagle
            Protection Act, 44
        Bureau of Land Management,
            43
        Clean Air Act, 43–44
        Clean Water Act, 43–45
        Comprehensive Environmental
            Response, Compensation
            and Liability Act, 43, 45
        Endangered Species Act, 43
        Federal Lands Policy
            Management Act, 43
        Federal Oil and Gas Royalty
            Management Act, 43
        Migratory Bird Treaty Act, 44
        Mineral Leasing Act, 42–43
        Oil Pollution Act, 43
        Outer Continental Shelf Lands
            Act, 42
        Reserved authority of states,
            45–46
        Safe Drinking Water Act, 43,
            45–46
        Submerged Lands Act, 42
Submerged Lands Act, 42, 57–58
Surface destruction test, onshore
    mineral ownership and leasing,
    47–48
Transfer or reservation of minerals,
    47
Unconventional onshore mineral
  extraction
        Generally, 61–72
        Hydraulic Fracturing, this
            index
United Nations Convention on the
    Law of the Sea, 58
Unitization, onshore mineral
    ownership and leasing, 52

OIL POLLUTION ACT
Environmental regulations, 105, 107
Oil and gas extraction, 43

OPEN-ACCESS SAME-TIME INFORMATION SYSTEM
Transportation of energy, 75

OPEN ACCESS TRANSMISSION TARIFFS
Electricity markets, federal
    regulation of, 176

ORGANIZATION OF ARAB PETROLEUM EXPORTING COUNTRIES
Imports and exports, 197

OUTER CONTINENTAL SHELF LANDS ACT
Generally, 4
Oil and gas extraction, 42, 57–59

PARTNERSHIP-FLIPS
Financing of renewable energy
    projects, utility-scale
    infrastructure, 159–160

INDEX

PHYSICAL SECURITY RELIABILITY STANDARD
Electric grid, cybersecurity, 89

PIPELINE AND HAZARDOUS MATERIALS SAFETY ADMINISTRATION
Transportation of energy, 91–96

PIPELINES
Markets, this index
Transportation of energy, safety, 90–96

PIPELINE SAFETY, REGULATORY CERTAINTY AND JOB CREATION ACT
Transportation of energy, 92

POOLING CLAUSES
Oil and gas extraction, onshore mineral ownership and leasing, 51–52

POWER PURCHASE AGREEMENTS
Electricity markets, federal regulation of, 168

PREDICTABILITY
Transportation of energy, interconnection of electric transmission lines, 81–82

PRODUCTION TAX CREDIT
Renewable energy projects, 143–145

PROPERTY ASSESSED CLEAN ENERGY
Financing of renewable energy projects, distributed infrastructure, 162–163

PRORATIONING
Oil and gas extraction, onshore mineral ownership and leasing, 52

PUBLIC UTILITY HOLDING COMPANY ACT
Electricity markets, federal regulation of, 171, 179

PUBLIC UTILITY REGULATORY POLICIES ACT
Electricity markets, federal regulation of, 178–179

PURPOSE OF ENERGY LAW
Generally, 3

QUALIFYING SMALL POWER PRODUCTION FACILITIES
Electricity markets, federal regulation of, 178–179

QUEUE MANAGEMENT
Transportation of energy, interconnection of electric transmission lines, 81–83

RAILROADS
Oil markets, federal regulation of, 186
Siting of coal transportation facilities, 34–35
Transportation of energy, safety, 94–96

RATES
Markets, this index

REAL ESTATE INVESTMENT TRUSTS
Financing of renewable energy projects, utility-scale infrastructure, 161

REGIONAL ENTITIES
Environmental Regulations, this index
Hydraulic fracturing, 69
Transportation of energy, North American Electric Reliability Corporation, 87–88

REGIONAL TRANSMISSION ORGANIZATIONS
Generally, 7
Electricity markets, federal regulation of, 168, 177–178, 180
Transportation of energy, 75–77, 79–81

RELIABILITY COORDINATORS
Transportation of energy, North American Electric Reliability Corporation, 88

RENEWABLE ENERGY
Generally, 3
Biofuels, this index
Environmental regulations, 114–117
Ethanol, this index
Hydropower, 134
Mandates, 140–150, 155–157
Renewable energy credits, 150–152
Renewable portfolio standards, 128, 149–150
Smart Grid and Distributed Generation, this index
Solar photovoltaic panels, siting of generation facilities, 14–15
Tax Incentives, this index

RESOURCE CONSERVATION AND RECOVERY ACT
Oil and gas operations, environmental regulations, 106

REVIEW
Appeal and Review, this index

RIVERS AND HARBORS ACT
Coal transportation, 35

ROYALTIES
Oil and gas extraction, onshore mineral ownership and leasing, 48

SAFE DRINKING WATER ACT
Hydraulic fracturing, 65, 107
Oil and gas extraction and operations, 43, 45–46, 105, 107

SAFETY
Transportation of Energy, this index

SAN BRUNO
Pipeline explosion, 92

SITING OF ENERGY FACILITIES
Generally, 13–39
Certificate of need or certificate of public convenience and necessity, electric transmission lines, 18
Coal transportation
    Generally, 32–39
    Export trends, 32–34
    Rail transportation, 34–35
Electric transmission lines
    Generally, 18–22
    Certificate of need or certificate of public convenience and necessity, 18
    Federal regulation, 20
    Load centers, 18–19
    Public utility commissions, 18
Eminent domain, oil pipelines, 23–24
Export trends, coal transportation, 32–34
Fuel abundance and fuel delivery infrastructure, generation facilities, 14
Generation facilities
    Generally, 13–18
    Fuel abundance and fuel delivery infrastructure, 14
    Hybrid approach, state and local regulations, 17–18
    Nuclear energy, 15
    Solar photovoltaic panels, 14–15
    State and local regulations, 15–18
    Thermal plants, 14
    Transmission lines, 14–15
    Utility-scale v. distributed generation v. community-scale, 13
    Water sources, 14
    Wind towers and turbines, 14–15
Hybrid approach to generation facilities, state and local regulations, 17–18
LNG terminals, 28–32
Load centers, electric transmission lines, 18–19
Natural gas pipelines, 27–28
Nuclear energy, generation facilities, 15
Oil pipelines
    Generally, 22–27
    Eminent domain, 23–24
Public utility commissions, electric transmission lines, 18
Rail transportation, coal, 34–35
Solar photovoltaic panels, generation facilities, 14–15
Thermal plants, generation facilities, 14
Transmission lines, generation facilities, 14–15
Utility-scale v. distributed generation v. community-scale generation facilities, 13
Water sources, generation facilities, 14
Wind towers and turbines
    Energy transmission lines, 18–20
    Generation facilities, 14–15

SMART GRID AND DISTRIBUTED GENERATION
Generally, 4, 209–231
American Recovery and Reinvestment Act, 215
Battery technology, electric vehicles, 222
Benefits of smart grid, 211–212
Charging stations, electric vehicles, 222
Data privacy, 217–218
Data standardization, 216–217
Electric vehicles
    Generally, 210, 220–225
    Battery technology, 222
    Charging stations, 222
    Lifecycle analysis of environmental impacts, 222–225
    Tax incentives, 221–222
Energy efficiency resource standards, 219
Energy Independence and Security Act, 214

Energy Policy Act, 213
Financing of Renewable Energy
 Projects, this index
Integrated resource planning, 219
Large-scale batteries, 225–226
Lifecycle analysis of environmental
 impacts, electric vehicles, 222–
 225
Micro grids, 210, 228–231
Smart grid interoperability
 framework, 214–215
Smart meters, 212
Storage of energy
 Generally, 225–228
 Large-scale batteries, 225–226
 Utility-scale generation v.
  distribution or microgrid
  generation, 226–227
Tax incentives, electric vehicles, 221–
 222
Utility-scale generation v.
 distribution or microgrid
 generation, storage of energy,
 226–227

SOLAR PHOTOVOLTAIC PANELS
Siting of generation facilities, 14–15

SOUTHWESTERN POWER ADMINISTRATION
Energy transmission lines, siting, 20–21

STORAGE OF ENERGY
Smart Grid and Distributed
 Generation, this index

SUBMERGED LANDS ACT
Oil and gas extraction, 42, 57–58

SULFUR DIOXIDE EMISSIONS
Electricity, environmental
 regulations, 120–121

SURFACE DESTRUCTION TEST
Oil and gas extraction, onshore
 mineral ownership and leasing,
 47–48

SURFACE MINING CONTROL AND RECLAMATION ACT
Generally, 5
Environmental regulations, 98–104

SUSQUEHANNA RIVER BASIN COMMISSION
Generally, 7
Hydraulic fracturing, 69
Oil and gas operations,
 environmental regulations,
 108–109

TAX INCENTIVES
Generally, 139–149
Biofuels, 147
Cash grant program, renewable
 energy projects, 145–146
Clean renewable energy bonds, 146–
 147
Coal bed methane gas, 142
Department of Energy loan
 guarantee program, renewable
 energy projects, 146
Electric vehicles, 221–222
Energy production and use,
 generally, 139–149
Ethanol tax credits, 140–141
Extension of renewable energy tax
 credits, 147–148
Fossil fuel tax credits and incentives
 Generally, 141–143
 Coal bed methane gas, 142
 Hydraulic fracturing, 142
Investment tax credit, renewable
 energy projects, 145
Production tax credit, renewable
 energy projects, 143–145
Renewable energy projects
 Generally, 140–141,
  143–149
 Biofuels, 147
 Cash grant program, 145–146
 Clean renewable energy bonds,
  146–147
 Department of Energy loan
  guarantee program, 146
 Ethanol tax credits, 140–141
 Extension of renewable energy
  tax credits, 147–148
 Investment tax credit, 145
 Production tax credit, 143–145
 State, local government and
  utility subsidies and
  incentives, 148–149
State, local government and utility
 subsidies and incentives,
 renewable energy projects,
 148–149

TEXAS COMPETITIVE RENEWABLE ENERGY ZONE
Electric transmission lines, siting, 19–20

THERMAL PLANTS
Siting of generation facilities, 14

THIRD-PARTY OWNERSHIP
Financing of renewable energy
 projects, distributed
 infrastructure, 163–164

TORT CLAIMS
Oil and gas operations, environmental regulations, 109–110

TRANSPORTATION OF ENERGY
Generally, 73–96
Access to electric transmission lines, 74–79
Bonneville Power Administration, 83
Brownouts and blackouts, 86
Costs of electricity transmission, allocation of, 79–81
Electric grid, safety, 86–90
Environmental Regulations, this index
Federal Energy Regulatory Commission, generally, 73 et seq.
Federal Power Act, 73–74
Federal Railroad Administration, safety, 94–95
High-hazard flammable trains, 95–96
High thread urban areas, railroads, 96
Independent system operators, 75–77
Interconnection of electric transmission lines, 81–85
Midcontinent Independent System Operator, 81, 83–84
North American Electric Reliability Corporation, safety, 87–88
Open-Access Same-Time Information System, 75
Order 890, access to electric transmission lines, 77–78
Order 1000, access to electric transmission lines, 78–79
Orders 888 and 889, access to electric transmission lines, 74–75
Pipeline and Hazardous Materials Safety Administration, 91–96
Pipelines, safety, 90–96
Pipeline Safety, Regulatory Certainty and Job Creation Act, 92
Predictability, interconnection of electric transmission lines, 81–82
Queue management, interconnection of electric transmission lines, 81–83
Railroads, safety, 94–96
Regional entities, North American Electric Reliability Corporation, 87–88
Regional transmission organizations, 75–77, 79–81
Regulation of electricity transmission, generally, 73–85
Reliability coordinators, North American Electric Reliability Corporation, 88

Safety
    Generally, 85–96
    Brownouts and blackouts, 86
    Cybersecurity, this index
    Electric grid, 86–90
    Federal Railroad Administration, 94–95
    High-hazard flammable trains, 95–96
    High thread urban areas, railroads, 96
    North American Electric Reliability Corporation, 87–88
    Pipeline and Hazardous Materials Safety Administration, 91–96
    Pipelines, 90–96
    Pipeline Safety, Regulatory Certainty and Job Creation Act, 92
    Railroads, 94–96
    Regional entities, North American Electric Reliability Corporation, 87–88
    Reliability coordinators, North American Electric Reliability Corporation, 88

TRESPASS
Hydraulic fracturing, 70

UNITED NATIONS CONVENTION ON THE LAW OF THE SEA
Oil and gas extraction, 58

UNITIZATION
Oil and gas extraction, onshore mineral ownership and leasing, 52

WATER POLLUTION
Oil and gas operations, environmental regulations, 105

WATER SOURCES
Siting of generation facilities, 14

WESTERN AREA POWER ADMINISTRATION
Energy transmission lines, siting, 20

"WHEELING"
Electricity markets, federal regulation of, 174–176

WIND TOWERS AND TURBINES
Siting of Energy Facilities, this index